Technology:
A World History

C1961

The
New
Oxford
World
History

Technology:
A World History

Daniel R. Headrick

OXFORD
UNIVERSITY PRESS

2009

OXFORD
UNIVERSITY PRESS

Oxford University Press, Inc., publishes works that further
Oxford University's objective of excellence
in research, scholarship, and education.

Oxford   New York
Auckland   Cape Town   Dar es Salaam   Hong Kong   Karachi
Kuala Lumpur   Madrid   Melbourne   Mexico City   Nairobi
New Delhi   Shanghai   Taipei   Toronto

With offices in
Argentina   Austria   Brazil   Chile   Czech Republic   France   Greece
Guatemala   Hungary   Italy   Japan   Poland   Portugal   Singapore
South Korea   Switzerland   Thailand   Turkey   Ukraine   Vietnam

Published by Oxford University Press, Inc.
198 Madison Avenue, New York, New York 10016

www.oup.com

Oxford is a registered trademark of Oxford University Press

Library of Congress Cataloging-in-Publication Data
Headrick, Daniel R.
Technology : a world history / Daniel R. Headrick.
p.   cm. — (The new Oxford world history)
Includes bibliographical references and index.
ISBN 978-0-19-515648-5; 978-0-19-533821-8 (pbk.)
1. Technology—History.   I. Title.
T15.H42   2008
609—dc22   2008033426

Printed in the United States of America
on acid-free paper

Frontispiece: *A turbine at the Niagara Falls Power Company.*
Photography Collection, Miriam and Ira D. Wallach Division of Art, Prints and Photographs,
The New York Public Library, Astor, Lenox and Tilden Foundations

Contents

Editors' Preface

The history of humans and technology is a long one, going back millions of years to the use of stones as tools and to their fashioning into more efficient devices through skillful flaking. Ancient peoples discovered the use of fire as a survival technology, only much later devising increasingly complicated systems of water management for irrigation and later still for hydroelectric power and many other uses. As communications technology developed closer to our own times, it brought people into greater contact and made them more knowledgeable and cosmopolitan. Medical and agricultural technology improved life expectancy, especially in our modern era; artificial organs could replace dying ones, and chemical and nuclear medicines could stop diseases such as cancers in their tracks.

Although such technology appears to have an exclusively personal function, making life more pleasant and efficient, ambitious leaders of ancient and more recent times have commandeered technology to help them build states and to conquer other peoples. Aqueducts stretching for hundreds of miles and the building of ships for warfare and trade were among the technologies that allowed leaders of states to maintain and expand their power. Increasingly, the comparatively simple weaponry of Stone Age people gave way to more complex machinery for conquest and destruction, weaponry that has been put to ever more devastating use in the past century.

It is hardly surprising, then, that people have had ambivalent feelings about technology of all sorts—and not just about the sophisticated machines of our own day. Pliny the Elder in the first century CE praised iron for its ability to cut stone and fell trees: "But this metal serves also for war, murder and robbery," he wrote in *Natural History*, "and this I hold to be the most blameworthy product of the human mind." Critics have also charged technology with pollution and other devastating effects on the natural world. For all its ability to provide increasing ease for the world's inhabitants, the case for technology's drawbacks is a powerful one, showing the tensions produced by the universal human capacity to invent.

This book is part of the New Oxford World History, an innovative series that offers readers an informed, lively, and up-to-date history of the world and its people that represents a significant change from the "old" world history. Only a few years ago, world history generally amounted to a history of the West—Europe and the United States—with small amounts of information from the rest of the world. Some versions of the old world history drew attention to every part of the world *except* Europe and the United States. Readers of that kind of world history could get the impression that somehow the rest of the world was made up of exotic people who had strange customs and spoke difficult languages. Still another kind of old world history presented the story of areas or peoples of the world by focusing primarily on the achievements of great civilizations. One learned of great buildings, influential world religions, and mighty rulers but little of ordinary people or more general economic and social patterns. Interactions among the world's peoples were often told from only one perspective.

This series tells world history differently. First, it is comprehensive, covering all countries and regions of the world and investigating the total human experience—even those of "peoples without histories" living far from the great civilizations. "New" world historians thus share an interest in all of human history, even going back millions of years before there were written human records. A few new world histories even extend their focus to the entire universe, a "big history" perspective that dramatically shifts the beginning of the story back to the Big Bang. Some see the new global framework of world history today as viewing the world from the vantage point of the moon, as one scholar put it. We agree. But we also want to take a close-up view, analyzing and reconstructing the significant experiences of all of humanity.

This is not to say that everything that has happened everywhere and in all time periods can be recovered or is worth knowing, but there is much to be gained by considering both the separate and interrelated stories of different societies and cultures. Making these connections is still another crucial ingredient of the new world history. It emphasizes connectedness and interactions of all kinds—cultural, economic, political, religious, and social—involving peoples, places, and processes. It makes comparisons and finds similarities. Emphasizing both the comparisons and interactions is critical to developing a global framework that can deepen and broaden historical understanding, whether the focus is on a specific country or region or on the whole world.

The rise of the new world history as a discipline comes at an opportune time. The interest in world history in schools and among the general public is vast. We travel to one another's nations, converse and work with people around the world, and are changed by global events.

War and peace affect populations worldwide, as do economic conditions and the state of our environment, communications, and health and medicine. The New Oxford World History presents local histories in a global context and gives an overview of world events seen through the eyes of ordinary people. This combination of the local and the global further defines the new world history. Understanding the workings of global and local conditions in the past gives us tools for examining our own world and for envisioning the interconnected future that is in the making.

Bonnie G. Smith
Anand A. Yang

Technology:
A World History

CHAPTER I

Stone Age Technology

In a place called Laetoli, in Tanzania, a family—a male and a female carrying a child—once walked across some fine volcanic ash. Their footprints, covered with ashes, remained untouched for 3.5 million years until 1978, when the anthropologist Mary Leakey discovered them. They are the oldest known footprints of Australopithecines ("southern apes"), a species that lived in southern and eastern Africa between 4.5 and 2.5 million years ago.

From the fossils of skulls that anthropologists have found, we know that the brains of these apes were as large as those of chimpanzees, about one-third the size of human brains today. Like other apes, they had strong jaws and teeth, with which they scavenged meat left over by other carnivores, as well as vegetable matter and whatever small animals they could catch. They differed from other apes in several ways, however. They lived in open grasslands, not in forests. Unlike all other mammals, they were bipedal; that is, they could walk upright comfortably. Their hands had opposable thumbs, with which they could grasp things. Holding objects in their hands, they could walk without slowing down. We do not know whether they carried sticks or hides because such organic matter has long since disintegrated. We do know, however, that they carried rounded stone cobbles and large pebbles long distances from the rivers where they found them. In short, they used found objects as tools.

Humans are not the only creatures that use tools; chimpanzees, vultures, sea otters, even insects will sometimes pick up a twig or a stone to get at food. Only humans, however, could not survive without tools, and only humans have in turn been shaped by the tools they use. How we got there is a story that began millions of years ago.

The oldest deliberately made tools we know of, found in the Omo Valley of Ethiopia, date back 2.5 million years. They, too, were river cobbles but with pieces broken off to make crude choppers with a sharp edge, useful for chopping wood or breaking the bones of animals to get at the marrow. The flakes that broke off were also sharp enough to cut hide and meat.

Gradually, the ability to walk upright, to manipulate objects with their hands, and to manufacture tools transformed not only the way of life of the Australopithecines but their very nature and anatomy. After millions of years, they evolved into a different genus, to which anthropologists give the name *Homo* or *hominid*, from the Latin word for "man." We cannot say that creatures with large brains "invented" tools; rather, brains, other anatomical features, and tools evolved together to create these creatures, our ancestors.

Several species of *Homo* belonged in the genus hominid, all of them living in Africa between 2.5 and 1.8 million years ago. The best known is *Homo habilis*, "handy man." These creatures' brains were half again as large as those of the Australopithecines, though still only half the size of ours. The fact that they carried cobbles up to nine miles from the riverbeds where they were found shows that they could plan for the future, something no other apes could do. They used these cobbles as hammers and made choppers by removing flakes from both sides, an improvement over their predecessors' tools. We do not know what other tools they made or how dependent they were on their simple technology. We know, however, that they were well adapted to surviving on the open savannas of Africa, for their anatomies and their choppers remained virtually unchanged for almost a million years.

We know much more about the creatures called *Homo erectus* ("standing man") who replaced these early hominids around 1.8 million years ago. They had brains two-thirds the size of ours. Like modern humans, their jaws and teeth were smaller than those of *Homo habilis*, and their arms were shorter and their legs longer. They were much less adept at biting and chewing and at climbing trees than earlier species. They could not have survived without tools.

Their tools, however, were much more developed than their predecessors'. We call their stone tools hand axes, or bifaces, because they were carefully flaked on both sides to provide a fairly even and long-lasting cutting edge. Some had a sharp point, and others, called cleavers, had a straight edge. Hand axes and cleavers could weigh as much as five pounds. They were multipurpose tools used to skin and butcher animals, to scrape skins, and to carve wood. Evidently, these hand axes served them well, for they hardly changed for close to a million years.

More important, *Homo erectus* mastered fire, the only creatures to do so. Fire allowed them to protect themselves from predators, to frighten animals, to warm themselves in cold weather, and to roast meat, which they needed to do because their small teeth had difficulty chewing raw meat. Australopithecines and *Homo habilis* had hunted, but only small or weak animals; otherwise, they scavenged the leftovers of more powerful predators, such as the big cats. *Homo erectus*, in contrast, were big-game hunters. Working in teams, they were able to drive wooly mammoths, larger than elephants, into swamps where they could be killed with spears and stones. Like all hominids, much of their nutrition came from vegetable matter collected, probably by females.

Thanks to fire and their superior hunting skills, *Homo erectus* could live in temperate climates and could therefore migrate from tropical Africa to other continents. More than a million years ago, *Homo erectus* reached the Caucasus, northern China, and Java, and later Spain and France. But they could not survive in really cold climates, like northern Eurasia, nor could they cross bodies of water; therefore, they never reached Australia, the Americas, or the islands of the Pacific.

Several species of *Homo erectus* existed at the same time. Sometime between 150,000 and 100,000 years ago in Africa, one species, and possibly more than one, evolved into a more advanced creature with a brain as large and jaws and teeth as small as ours. We call this creature *Homo sapiens*, the "wise man," because we think of ourselves as wise. It was so similar to us that some anthropologists claim that if one of them reappeared on earth and sat on a bus seat next to us, we would think it was just another passenger.

There were two distinct species within the genus *Homo sapiens*: one with thick brows, strong bones, and the physique of a wrestler, called Neanderthal, after the Neander Valley in Germany where their remains were first found; the other, called archaic *Homo sapiens*, was physically exactly like modern humans. These two species lived side by side in the Levant (the lands bordering the eastern Mediterranean), but they did not interbreed. Then, 35,000 or 40,000 years ago, the Neanderthals vanished, and no one knows why.

The technology of the two species was identical and considerably more sophisticated than that of *Homo erectus*. Instead of making one kind of hand axe, they made many different stone tools for different purposes: stone spearpoints they attached to wooden shafts, blades of various sizes, and curved scrapers used to prepare hides, among others. Microscopic analysis shows that different tools were used to cut wood, to saw bones, to cut meat, and to scrape antlers.

Australopithecines, *Homo habilis*, and *Homo erectus* had distinctive stone tools but barely changed them for hundreds of thousands,

These stone points, dating from around 4200–3000 BCE, came from North Africa. Their concave bases indicate that they probably were used as arrowheads. Logan Museum of Anthropology, Beloit College

even millions, of years. Similarly, the tools of archaic *Homo sapiens* and Neanderthals changed very slowly over tens of thousands of years.

Then, 70,000 years ago, an explosion of innovations began, not only in tools but also in aspects of life unknown to previous hominids: art, religion, and ocean navigation. Some anthropologists call this event the Big Bang. Here was something new in the world: human culture, changing incomparably faster than the slow biological evolution of species.

We know much more about the material culture of modern humans than about that of their predecessors because modern humans created far more things and because many of the things they made out of organic matter—bones, antlers, hides, and wood—have survived over the past 70,000 years, especially in cold places where earlier creatures would not have ventured.

Consider just one kind of tool, the sharp-edged stone. Modern humans made a great variety of tools for cutting, scraping, and piercing, even burins or chisels used to engrave fine lines on antler and bone. They even made microliths, tiny pieces of sharp stone that they embedded in a bone or wooden haft to form a saw.

One anthropologist calculated how much cutting edge hominids were able to get from a one-pound piece of flint. *Homo habilis*, 2 million years ago, could break off a flake, leaving three inches of cutting edge; *Homo erectus*, 300,000 years ago, could obtain eight to twelve inches from the hand axe and the flakes; a Neanderthal, 100,000 years ago, could obtain 30 to 40 inches of cutting edge; by 30,000 years ago, a skilled hunter could turn that pound of flint into 30 to 40 *feet* of blades.

In 1991, mountain climbers found the complete remains of a man, frozen and perfectly preserved since around 3300 BCE, in a glacier in the high Alps along the border between Italy and Austria. The Ice Man, as he is now known, was dressed in a leather cap, vest, and leggings sewn with leather thongs. On his feet he wore calfskin shoes padded with grass for warmth. Over his clothes, he wore a cloak of woven grass. He carried the tools of a hunter: a bow, a quiver filled with flint-tipped arrows, a bag containing flint knives, scrapers, and burins for punching holes in leather. He even had an axe with a wooden handle and a copper blade, one of the very first metal tools. His equipment was just as sophisticated as that of much more recent hunters who need to survive in cold places, such as the Indians of the Rocky Mountains or the Inuit of the Arctic. His equipment was not only useful, but it was also dangerous, for he died of a wound from an arrowhead lodged in his shoulder. This was the first known case of one human killed by another.

As the Ice Man's equipment shows, humans made all kinds of useful things never seen before: they sewed clothes with bone needles, they wove ropes and nets, they carved fishhooks, and they made spears, spear throwers, and later, bows and arrows. Like their predecessors, they hunted big game and gathered nuts, fruits, and berries. But they also knew how to fish and catch shellfish and sea mammals. They made strings and ropes out of vegetable fibers and used them to make fishnets, fishing lines, and necklaces of beads.

With this equipment, humans ventured into ever-colder climates. In western Europe, much of which was still covered with ice, they lived in caves overlooking valleys through which great herds of animals migrated twice a year. To light their way into deep caverns, they made oil lamps. In the steppes of southern Russia and Ukraine, where there were no caves, they built houses out of the ribs of mammoths covered with hides, buildings large enough to shelter 50 people. They even knew how to store meat in pits covered with heavy stones to keep other animals away.

The artifacts of modern humans went far beyond the needs of survival. Whereas the artifacts of *Homo erectus* and Neanderthals were purely practical, humans made objects with no known practical application, which were created, instead, for religious, magical, or esthetic reasons. As early as 70,000 years ago, they made bone spearpoints smoother than was needed for hunting and even engraved them. They made musical instruments, such as a flute carved out of bone, 32,000 years ago. They sculpted figures of animals out of bone or ivory. Small sculptures of women show them wearing string skirts, some with metal beads at the ends. They used pigments and dyes to paint pictures of animals drawn with great artistic talent on the walls of caves, sometimes hundreds of yards underground. They carved stone spearpoints in the shape of a leaf so thin they could not possibly have been used for hunting. They decorated themselves with beads and perforated seashells and animals' teeth. They also buried their dead with ornate objects, like the 60-year-old man buried in Russia 28,000 years ago with pendants, bracelets, necklaces, and a tunic on which hundreds of ivory beads had been sewn. They had something new in the world: a sense of beauty.

What happened to transform archaic *Homo sapiens* into modern humans? Their bodies and brains were identical, so the change must have been purely cultural. Of their culture, we know only the material artifacts that have survived. So we have to make some educated guesses. The creation of objects that were not immediately practical or necessary for survival gives us a clue. These objects—cave paintings, musical instruments, sculptures, and adornments—are symbols that represent ideas such as beauty, control over animals, or life after death.

All humans today, and throughout historic times, express their ideas in language as well as in artifacts. Could it be possible that the sudden change in the creativity and life of *Homo sapiens* happened when they learned to talk? If so, then it explains why humans all over the world suddenly began to use symbolic representation and act in creative ways at around the same time: they learned language, symbols, and skills from one another. Cultural evolution, tied for millions of years to biological evolution, was now free to race ahead. Since the sudden emergence of symbolic representation, human culture has never slowed down or ceased to find new and more ingenious ways of doing things.

Like *Homo erectus* before them, *Homo sapiens* liked to travel. From their original homeland in Africa, they migrated to southwestern Asia 100,000 years ago. They reached South and Southeast Asia and Indonesia 70,000 years ago. By 40,000 years ago, they had settled in western Europe. Between 35,000 and 15,000 years ago, they occupied the steppes of southern Russia and Siberia, a more forbidding landscape.

They were now in territory no hominid or other primate had ever inhabited before. They reached New Guinea 40,000 years ago and Australia 5,000 years later, at the very latest, and possibly 10,000 or 15,000 years before that. At the time, the oceans were much lower than they are now, and these two great land masses were joined in a continent we call Sahul. But between Sahul and its nearest neighbor Sunda (which then included Asia and the Indonesian archipelago) stretched 62 miles of open water. In other words, to reach Sahul, the ancestors of today's New Guineans and Australian aborigines had to build boats large enough for several people, stock them with provisions for a journey of several days, and venture into the unknown. We have no idea what these craft were like, but the very fact that they reached Sahul attests to the ingenuity as well as the courage of these ancient mariners.

Just as mysterious and controversial is the arrival of the first humans in the Americas. Some anthropologists claim that humans arrived as far back as 45,000 ago, but most find their evidence unconvincing. All agree, however, that humans reached the New World 15,000 years ago at the latest and had spread to every corner of this continent within 3,000 years. They probably came on foot across the land bridge that then connected Siberia and Alaska, a forbidding land of glaciers and tundra. Then again, they may have paddled canoes along the coast, surviving on seafood and marine mammals; if they did, their campsites, and all the evidence thereof, are now under a hundred feet of ocean.

Like the first Australians, the first Americans were skilled hunters who made fluted stone spearheads called Clovis points after the town of Clovis, New Mexico, where these points were first discovered. The first Americans found their new homelands populated by many species of large animals we call megafauna: mastodons, mammoths, giant sloths, camels, bison, and moose. Most of these huge animals became extinct just around the time that humans appeared. Perhaps it was not a coincidence.

By 10,000 years ago, humans occupied almost every piece of land on earth except for the Arctic, Antarctica, and the islands of the Pacific. Their tools and artifacts were becoming ever more elaborate. So was their diet, as they hunted more kinds of animals, fished more efficiently, and gathered a greater variety of plant foods. To do so, they needed not only a more complex toolkit but also a deeper understanding of plants and animals, their behavior, and their value to humans. The technology of Stone Age people may seem simple to us, but their knowledge of their natural environment must have been enormous and has perhaps never been surpassed.

Of all the earth's environments, the shores of the Arctic Ocean offer the greatest challenges to human life. Nowadays, those who venture there from the temperate zone must bring everything they need to survive: ships, airplanes, snowmobiles, and tons of supplies and equipment. Yet long ago, the Inuit mastered the Arctic without elaborate imported paraphernalia.

The first inhabitants on the American side of the Arctic Ocean, ancestors of the Inuit, came from Siberia 10,000 years ago to hunt caribou. Later inhabitants ventured out onto the ice to kill seals as they came up to breathe at air holes. Their descendants went out in teams to kill whales that approached the shore. On land, they hunted with bows and arrows, but to kill whales and sea mammals, they fashioned harpoons with detachable heads attached to sealskin floats by lines of sinew. Once they impaled an animal, they could track its underwater movements by following the floats on the surface. Their boats—*kayaks* with which to hunt seals and larger *umiaks* to carry several people and to hunt whales and walruses—were made of wood and animal skins. To travel on snow, they used sleds pulled by dogs. Their houses were built of stones and sod; in the winter out on the ice, they built igloos of snow. For heat and light in the months-long winter nights, they made lamps in which they burned animal fat and whale blubber. Their clothes and shoes were made of the skins of seals, polar bears, and other animals. Having the right clothing was a matter of life or death in the Arctic. Making them was the work of Inuit women, who spent hours softening the hides by chewing them and sewing them into airtight and watertight clothing for each member of the family.

To survive, hunter-gatherers had to keep moving, following the herds of animals or seeking places where plants were ripening. Seldom did they settle in one place for long, for very few places on earth could support a group of foragers year round. One such place was Mount Carmel in Palestine, where a people we call the Natufians hunted with bows and arrows, fished with hooks and harpoons, and collected berries, fruits, nuts, and other edible plants. They reaped wild grains with bone sickles into which they inserted small flint teeth and then ground the grains with millstones. In Syria, foragers built a permanent village of 300 to 400 inhabitants we call Abu Hureya, where they lived by hunting and gathering for more than 2,000 years.

Another place foragers settled year round was in southern Japan, where a warm, rainy climate and the close proximity of forests, mountains, and seacoast provided a diversity of wild foods year round. There, 12,000 years ago, foragers began making pottery, thousands of years before anyone else in the world. Their first pots were large cone-shaped earthenware cooking pots, clearly too heavy to carry around. Later, they decorated their pots by pressing ropes into the soft clay before firing them, giving the pots, and the people who made them, the name *Jomon* (Japanese for "rope coil"). These people built villages of 50 or more dwellings and buried their dead in cemeteries. Their numbers may have reached a quarter million, with the highest population density and possibly the highest standard of living of foragers anywhere in the world. Their culture lasted for 10,000 years, long after their neighbors in Korea and China had developed an entirely different way of life.

Very few humans were as lucky as the Jomon people. In most places, necessity forced Stone Age hunter-gatherers to shift from hunting large animals to foraging for an ever-greater variety of wild foods. A slowly growing population needed more food, but previous migrations meant there were fewer places not already inhabited by other humans. With more mouths to feed and nowhere to go, people had to intensify their local foraging or starve. As good hunting grounds became crowded, hunters clashed more often; their skeletons, like the Ice Man's, show marks of violence. There was another alternative, however: helping edible plants grow and raising captured animals. This happened quite independently in several places around the world, proving that food production was not the "invention" of some lone

The first potters were the Jomon people of southern Japan, who pressed rope into clay to create elaborate decorations, such as this jug's ornate handles. Metropolitan Museum of Art

genius but a necessity that many people responded to in a similar fashion. Technological innovation, in this case growing plants and raising animals, was more of a change in culture and a new way of life than a new set of tools and artifacts.

The first place people began growing food was the Middle East, specifically a region called the "Fertile Crescent" that stretches north along the Mediterranean from Palestine through Syria and then southeast into the hills of Iran that overlook the Tigris-Euphrates Valley. From 12,000 to 14,000 years ago, with the end of the Ice Age, the climate of this region became warmer, and grasslands expanded. Among the grasses that grew in the region were several that produced edible seeds, in particular wild wheat and barley. At first, people harvested only the wild seeds, but around 12,000 years ago, they began to sow some seeds in favored locations and then remove competing plants and water the growing crop. In short, they began to garden. Gardening could be interrupted and picked up again (unlike hunting), and it could be combined with the nursing of babies and the care of small children. At a burial site in Abu Hureya in Syria, dating to 9700 BCE, the bones of women (but not of men) show malformations of the toes, knees, and vertebrae due, probably, to hours spent grinding grain on a grindstone. Women then, as later, prepared the food.

The transition from planting a few seeds to supplement a diet of wild foods to depending largely on domesticated plants took 2,000 years or more. To obtain more food with less effort, these early gardeners had to select seeds through a process of trial and error. They cleared land, sowed seeds, weeded, watered, harvested their crops, and generally adapted their activities to the cycle of plant growth. They also changed their way of life, settled down in villages, and made pots in which to cook their food. Not everyone preferred such labors to the wandering life of hunting and gathering. As one twentieth-century hunter-gatherer told a visiting anthropologist, "Why should we plant when there are so many mongongo nuts in the world?"

As people settled down to growing plants, they also began domesticating animals. The first animals to be domesticated were probably dogs that hung around the camps and followed humans around, waiting for scraps. Nomadic hunters also understood the behavior of animals like wild sheep and goats, whose herds they followed. When they caught young animals, rather than eating them right away, they penned them in and fed them to be eaten later. They let the more submissive ones breed, thereby producing, after hundreds of generations, tame animals that would not flee or fight approaching humans. Sheep and goats were the first herd animals to be domesticated, followed by pigs and donkeys and much later by cattle.

Domesticated animals offered a valuable addition to the lives of early farmers, compensating them for the disadvantages of relying on domesticated plants alone. They could be slaughtered and eaten at any time, not just at the end of a successful hunt or after the harvest like vegetables. After centuries of breeding goats, sheep, and cattle, humans learned to milk them; in most societies, milking cows, churning butter, and making cheese and yogurt were the work of women. Sheep bred for a soft fleece provided wool for clothing. And animal droppings were used as fertilizer or fuel.

Animal husbandry was not confined to mixed farming communities. Sheep and goats were herded to distant pastures according to the seasons, up into the mountains in the summer and down to the valleys in winter. After the domestication of horses around 1000 BCE, herding peoples kept their flocks of sheep and herds of cattle on the move year round, thereby inaugurating a new way of life, pastoral nomadism, that contrasted with, and sometimes threatened, the settled lives of farmers. When cattle were domesticated, many peoples in dry areas like central Asia and the Sudanic belt across Africa became full-time herdsmen.

The transition from foraging to farming took place in several centers around the world. The inhabitants of the Yellow River Valley in northern China began growing millet around 6500 BCE and later added soybeans, sorghum, and hemp. By 6000 BCE and possibly earlier, people in southern China and Southeast Asia cultivated rice. Taro and bananas also originated from that region, as did pigs, chickens, and water buffalo. From southern China, rice cultivation spread to northern China and Korea and, by 400 BCE, to Japan as well.

A similar diffusion took place from the Levant to Egypt and Europe. The peoples of Egypt and Greece, where the climate was similar to that of the Levant, adopted wheat and barley by 6000 BCE. Central and western Europe lagged because these Middle Eastern crops did poorly in colder, wetter climates; not until oats and rye were domesticated could the inhabitants rely on crops for most of their food.

The idea of horticulture may have spread from Egypt to sub-Saharan Africa, but Middle Eastern crops could not survive in the tropics. There, horticulture had to await the domestication of local plants, such as millet and sorghum and, in the moister regions, yams. An African variety of rice was first grown along the Niger River around 3500 BCE and later spread to Guinea and Senegambia. Not until the first millennium BCE could horticulture feed substantial numbers of people. Tending a garden could be done while caring for small children. For many centuries, this was the work of women, while men hunted or herded large animals.

The peoples of the Americas were cut off from the Eastern Hemisphere, so their domestication of plants and animals was completely independent of the rest of the world. The environment of the Americas also presented greater challenges than that of the Eastern Hemisphere, for there were fewer wild animals that could be tamed and the wild plants, while numerous, were difficult to domesticate. As a result, the transition to full food production took much longer than in the Middle East or East Asia. Around 5000 BCE, the people of central Mexico began experimenting with teosinte, the ancestor of maize, later adding squash, beans, tomatoes, and chili peppers. Yet it was not until 1500 BCE that most of their food intake was from farming. The only animals they domesticated were dogs, ducks, and guinea pigs.

The Indians of South America created an entirely different form of horticulture based on potatoes, quinoa (a grain), and beans. While the farmers of the Middle East learned to use donkeys, cows, and oxen to pull plows, these large animals did not exist in the Americas. The llamas and alpacas that South American Indians domesticated could be used as pack animals and raised for their meat and wool, but they were too small to be ridden or made to pull a plow. Without stronger animals, farming was much more heavily dependent on human labor than in the Eastern Hemisphere, imposing limits on the Indians' diets and productivity.

The gradual development of horticulture and agriculture transformed the world. People who grew or raised their own food could obtain much more from a given piece of land than foragers ever could. Fertile land could support up to a hundred times more farmers than foragers. As their numbers increased, farmers migrated to areas inhabited by hunters and gatherers, whom they soon outnumbered and, in many places, replaced entirely. The cultures and ways of life of hunters and gatherers, which had lasted for millions of years, only survived in places too dry to farm, like the plains of North America and central Asia, or too humid, like the rain forest of equatorial Africa. To obtain lands for their fields and ashes to fertilize their crops, farmers cut down forests and burned the trees. This transformed the global environment much more rapidly than ever before, although still slowly compared to our own times.

One major result of the agricultural revolution was the proliferation of settlements, villages where people lived year round, while they farmed and continued to forage in the vicinity. Jericho, in Palestine, may have been the first such village, dating back to 10,500 BCE. Its inhabitants grew wheat, barley, lentils, and peas. Within 1,000 years, the inhabitants were getting 80 percent of their food from growing barley and wheat and raising goats, and the rest was from foraging. They traded local products for obsidian, a volcanic glass with a sharp cutting edge used to harvest grain. They also imported seashells and turquoise from far away to make jewelry.

For protection against marauders, they surrounded their town with a wall and a ditch. By 7350 BCE, Jericho had grown into a town of some 2,000 inhabitants. Jarmo, in northern Iraq, was another such village, founded around 7000 BCE, that sheltered about 150 inhabitants. Inside the walls, the inhabitants built small houses of sun-dried bricks with flat roofs and less substantial huts out of reeds. In other places with a wetter climate, farmers often built houses out of wood with thatched roofs.

Settling down in one place opened up a world of possibilities. Villagers could devote time to creating new technologies; more important, they could make objects too heavy or fragile for people to carry from

A brick wall protected the city of Jericho from invasion. Originally, this mud-brick wall was on top of a stone outer wall; only the stone part is still intact. Library of Congress LC-M32-290 [P&P]

place to place. Archaeologists call this early agricultural period the Neolithic, or New Stone Age, because the stone tools dating from this period were polished smooth. Many of them were axe and adze heads, and polishing made them less prone to fractures. But this was only one of many technological changes that characterized this period.

The tools that farmers needed were very different from those of foragers. Axes were used to chop down trees. Once cleared of trees, the land had to be prepared with digging sticks and stone-bladed hoes. Farmers harvested grain with bone or wooden sickles into which they inserted sharp stones. To store grain, sometimes for as long as a year, until the next harvest, they made baskets out of reeds and bins of wattle and daub (branches covered with mud). Protecting grain from insects and rodents demanded even tighter containers, as did carrying water and cooking. Clay pottery, long known to the Jomon people of Japan, appeared in the Middle East between 7000 and 6000 BCE, soon after the first farming villages. Before grain could be eaten, it had to be crushed with grinding stones or with wooden mortars and pestles. People also cooked by dropping hot stones into a water-filled clay pot, causing the water to boil. Their diet contained much more starch and less protein than that of hunters and gatherers. Their remains show that they were smaller and less healthy than their ancestors.

Textiles also appeared in the early farming villages, replacing animal skins. The first woven cloth we know of was made in Jarmo. Making cloth or rugs required a source of fibers, such as hemp, flax, cotton plants, or wool from sheep. The fibers first had to be cleaned, carded with combs to make them lie parallel, and spun on spindles to make yarn. They were then woven on looms, the first of which dates from around 6000 BCE. The sources of fibers and the equipment needed to make cloth required year-round settlements. Like tending a garden, spinning yarn and weaving cloth were tasks that could be combined with the care of small children. For that reason, this work was almost universally done by women. The earliest representations of spindles and of looms from Egypt, Mesopotamia, India, and Scandinavia show women spinning and weaving.

Neolithic people, by definition, made tools out of stone. They used not only local stones but also fine-grained stones—flint, chert, and obsidian—imported, in some cases, from hundreds of miles away. This meant they had extensive trading networks—for instance, between Mesopotamia and Spain or Britain or between the Gulf of Mexico and the Great Lakes of North America. They also used metal when they found it in the raw state, as occurred in deposits of copper, silver, or gold; in the Americas, all metals were of this variety. The Ice Man of the Alps carried a copper-bladed axe. Smelting—that is, making copper from ores—began in Anatolia by 5000 BCE, in China before 2800 BCE, in the Andes around 2000 BCE, and in West Africa by 900 BCE.

The most extraordinary expression of Neolithic technology is the construction of megaliths, or huge stone monuments. Neolithic people erected monuments in many places, from western Europe to Easter Island in the Pacific. The earliest were the temples of Malta, built 6,000 years ago. Most astonishing of all, however, is the great stone circle at Stonehenge in England. The largest stones weigh between 25 and 50 tons each and were transported 25 miles overland. Eighty-two blue stones weighing five tons apiece were brought from 150 miles away, partly on rafts. All of the work was done without the help of pulleys or other devices. Archaeologists estimate that building Stonehenge required the work of several thousand people over the course of many years between 2800 and 1100 BCE. Yet on certain days of the year, the stones are perfectly aligned with the moon and the sun, leading many to speculate that they were used for astronomical observation or to establish a calendar.

During the millions of years between the Australopithecines and the builders of Stonehenge, the ancestors of today's human beings underwent a long evolution, both biological and cultural. Beginning with the Australopithecines, apes who could walk on their hind legs, hominids changed over time, until the most recent of their species, *Homo sapiens*, appeared more than 100,000 years ago.

Because they walked upright, hominids could use their hands to carry and make things. For several million years, what they did with their hands was limited by their brains. Thus, their technology changed as slowly as their brains and bodies. In the last 100,000 years, however, the pace of change has accelerated. Although the bodies and brains of modern humans have hardly changed in this time, the variety, effectiveness, and meaning of the artifacts they created have increased exponentially. Thanks to their techniques and artifacts, by 10,000 BCE, humans had become the most efficient foragers and the most successful predators the world had ever seen, and they had spread to almost every region of the world. *Homo sapiens* had become the dominant creature within the natural world.

At this point, living within the natural world no longer sufficed. Propelled perhaps by a growing population or a changing climate, humans found ways to manipulate the natural world to their advantage.

A revolutionary development, the production of food through the domestication of plants and animals, allowed people to settle down and create many other technological artifacts from cloth to megaliths.

But was this progress? The bones of early farmers that archaeologists have studied show that they were shorter, more poorly nourished, and more disease ridden than their hunter-gatherer ancestors. They must also have worked much harder during certain seasons of the year than foragers ever did. Most of their efforts went to ensuring their survival, a risky venture when a drought, a flood, or a plague of locusts could destroy their livelihood. Yet there was no turning back. Farming and herding could support more people in a given environment than foraging ever could. As their numbers multiplied, agricultural people could not go back to hunting and gathering but could only practice their new way of life ever more intensively. Some were more successful than others.

CHAPTER 2

Hydraulic Civilizations (4000–1500 BCE)

The Book of Genesis in the Bible describes the third day of the Creation in these words:

God said, "Let the water below the sky be gathered into one area, that the dry land may appear." And it was so. God called the dry land Earth, and the gathering of waters He called Seas. And God saw that this was good. And God said, "Let the earth sprout vegetation: seed-bearing plants, fruit trees of every kind on earth that bear fruit with the seed in it." And it was so. The earth brought forth vegetation: seed-bearing plants of every kind, and trees of every kind bearing fruit with the seed in it. And God saw that this was good.

We now know how this happened. Six thousand years ago, a people called Sumerians began separating land from water and planting crops in the newly reclaimed wetlands rather than relying on rainwater as Neolithic farmers had done. In doing so, they created the first civilization.

The word *civilization*, as historians and anthropologists use it, refers to large-scale societies whose members contribute taxes, labor, or tribute to the state and pay homage to their leaders. Such societies were radically different from Neolithic villages or foraging bands, whose members knew each other and were related by blood or marriage. Not only did civilizations include far more people, but they also built monuments and cities, invented writing, mathematics, and calendars, and created elaborate religions, literatures, philosophies, and other forms of culture. Some civilizations eventually collapsed or were conquered by outsiders, but others survived for millennia. In later centuries, people often looked back nostalgically to a "Golden Age" or a "Garden of Eden" before they became civilized. But once they had crossed the line, they could never return.

Unlike Neolithic villages where everyone helped provide food, in larger societies, some people performed tasks other than farming or herding. A few were full-time religious, political, or military leaders. Some were warriors, artisans, and merchants. And others were servants to the elites or upper classes. To feed them, the farmers, herdsmen, and fishermen had to produce more food than they themselves consumed. The key to the transformation from Neolithic villages to civilizations, therefore, was the methods used to produce a surplus of food to feed those who did not farm. New and more productive farming practices went hand in hand with a radically new organization of society.

The earliest civilizations did not arise in fertile rain-watered lands in the temperate zone. Instead, they began in dry or desert regions where water came from a river, a lake, or a swamp. Farmers who grew crops on the very banks of the river or the shores of the lake or swamp were always at the mercy of devastating floods or droughts. When they succeeded in controlling the water, however, the results were spectacular. Whereas Neolithic farmers in the Middle East might hope to reap four or five grains of barley for every grain they planted on rain-watered land, in a river valley, a grain of barley receiving the right amount of water during the growing season could yield up to forty grains.

The farmers who settled closest to the rivers could depend on periodic floods to water their fields. Those who came later, however, settled further from the riverbanks. To bring water to their fields, they had to dig canals, dikes, and other earthworks. Building and maintaining these works required the labor of hundreds, even thousands, of men directed by a cadre of supervisors. Although farmers had to contribute their labor, they were not slaves driven by men with whips. People obeyed because they realized the need to work together, because of the peer pressure of their neighbors, and because they were afraid that refusing would bring down the wrath of the gods. Moreover, they knew that they had nowhere else to go. In rain-watered environments, people could wander off seeking new land, but in desert regions, survival was impossible outside the river valleys.

The place where the first civilization arose was Iraq, a land the Greeks called *Mesopotamia*, the "land between the rivers" Tigris and Euphrates. The valley has good alluvial soil but is difficult to farm. It is very hot and dry in the summer and cold and dry in the winter. Although little rain reaches the valley, in the spring water rushes down from the mountains to the east and north when the snows melt. The rivers carry a great deal of silt that gradually raises them above the surrounding plains until they overflow their banks in devastating floods. All the peoples of the region told legends of the flood, most famously the Hebrew story of Noah's Ark told in the Bible (Genesis 5–9).

To the Neolithic farmers who lived in the surrounding hills, the flood plain presented both an opportunity and a challenge. By the sixth millennium BCE, the bolder ones were moving down into the plains and building villages. By the fifth millennium, they were digging short feeder canals to irrigate their fields and drain excess water. To keep the floods from washing away their crops, they built dikes. To hold some of the water back when the floods subsided in the summer when the crops needed water the most, farmers built small reservoirs. Keeping the water flowing was a constant task because silt clogged the canals and the salt and gypsum it contained would poison the fields if they were not properly drained. As the population grew, farmers drained marshes and built canals and reservoirs ever farther from the rivers, requiring ever larger work crews. Success depended on good leadership and the cooperative work of thousands.

By carefully watering the rich alluvial soil, farmers grew an abundance of barley, wheat, and date palms, along with lentils, beans, peas, onions, and reeds, out of which they built houses and boats. They raised sheep, goats, donkeys, cattle, and pigs and caught fish in the canals. There was more than enough for the farmers and herders to eat. After 3500 BCE, villages in the wetlands of southern Iraq grew into towns, and towns grew into cities. The techniques used by the Sumerians gradually spread up the rivers and to the outer edges of the valley. After 2000 BCE, farmers began watering their fields with a *shaduf*, or "well-sweep," a long pole with a bucket at one end and a counterweight at the other. Instead of using a hoe or a digging stick as their ancestors had, they cultivated their fields with an ox-drawn plow and planted seeds with a seed drill, a device that dropped seeds at regular intervals. This shift from horticulture to true agriculture produced much greater yields. Under the direction of their rulers, gangs of laborers dug canals up to 75 feet wide and many miles in length. The most famous of their kings, the lawgiver Hammurabi who reigned from 1792 to 1750 BCE, named one of his canals "Hammurabi-spells-abundance."

Egypt was an easy land to farm compared with Mesopotamia. The Nile flooded its valley in late summer and early fall, after the harvest. Unlike the Tigris and Euphrates, the timing of the Nile flood was predictable, and the silt its waters carried was fertile and salt-free. The Egyptians built low dikes that divided the land into basins, letting water stand for about a month to deposit its silt and soak the soil before it was allowed to flow downstream to the delta of the Nile. Crops were planted in October or November and harvested in April or May, before the next flood.

Neolithic peoples had inhabited the Nile Valley for centuries, farming on the riverbanks and hunting and fishing the wild game in which the land abounded. In the fourth millennium BCE, Egypt was divided into little kingdoms, each of which had a "water house" that planned the building of dikes and the soaking of the fields. In the early third millennium, after lower and upper Egypt were united under the Pharaoh Menes, engineers installed what we call nilometers, devices that measured the height of the river. The regularity of the floods led them to devise a 365¼-day calendar. When they saw Sirius, the brightest star, rising in the dawn sky in line with the rising sun, they knew the flood was imminent. They also developed surveying instruments and a practical geometry to help them place boundary stones to mark the edges of fields and irrigation basins. They used shadufs and other devices such as pulleys and treadmills to lift water above the level of canals. The resulting food surpluses not only supported the creation of the elaborate culture and awe-inspiring monuments for which ancient Egypt has always been famous, but they also produced the most secure and sustainable civilization the world has ever known—one that lasted, with only brief interruptions, for 3,000 years.

Thirteen hundred miles east of Mesopotamia, the Indus River flows through Sind, now a province of Pakistan. The environment of the Indus Valley was similar to that of Mesopotamia, with a rich soil, a hot, dry climate, and a violent river that periodically flooded the plain. Unfortunately, we know far less about the civilization that arose there than about Sumer or Egypt because the few writings that have survived have not yet been deciphered. We know that the organization of flood control in the valley began between 3200 and 2600 BCE. Villagers dug irrigation and drainage canals and built embankments to control the floods and protect their settlements. They grew wheat and barley and traded these crops with nearby nomadic tribes for metals, semiprecious stones, timber, sheep, and goats. They also traded with the peoples of Sumer and the Arabian Peninsula, as evidenced by pieces of Indus pottery and metal objects found in both places. Some time after 1700 BCE, for reasons we do not fully understand, the population shrank, water control was abandoned, and the cities of the Indus Valley were destroyed by floods.

The distinctive cultures of Egypt and the Indus Valley were inspired by the example of nearby Mesopotamia. In China, Mexico, and Peru, three different agricultural systems developed quite independently of outside influences. The earliest center of civilization in East Asia appeared on the plains of northeastern China, along the

Yellow River. The land there was exceptionally fertile, composed of *loess*, windblown and waterborne silt that was soft enough to cultivate with digging sticks. On average, rainfall was adequate for agriculture, unlike the river valleys of Mesopotamia, the Nile, and the Indus, and farmers could plant dry-land crops such as millet and wheat. Some years, however, drought parched the land. Worse were the years when too much rain fell on the mountains of central Asia. Then the Yellow River became so laden with silt (hence its name) that it built up its bed above the flood plain and then broke through its natural embankments in raging floods that swept everything in their path. That is why the Chinese people call it "China's sorrow."

By the fourth millennium, Neolithic farmers were clearing the forests and building dikes, channels, and reservoirs to control the waters of the Yellow River. But to protect the inhabitants and support a growing population, better flood control was needed. King Yu, founder of the legendary Xia dynasty, is credited with the first large-scale flood-control project in China, around the year 2200 BCE. During the Shang dynasty (ca. 1600 to ca. 1046 BCE), the first one for which we have evidence in the form of pot shards, walls, and other remains, the Yellow River plain was dotted with thousands of villages whose inhabitants grew millet and wheat, raised pigs and silkworms, and made pottery. Above them ruled an aristocracy of warriors who supervised the engineering projects, built cities, and fought with their neighbors.

If China was almost cut off from other early civilizations, the Americas were completely isolated. Thus, the Native Americans proceeded at their own pace, undisturbed by outside influences until Columbus arrived in 1492. On their own, albeit much later, they created impressive civilizations similar in many ways to those of the Old World, based on water control in similar environments.

As in the Old World, ecological conditions varied from one part of the Americas to another, and so did the methods people devised to make best use of the land and the water. Six thousand years ago, the inhabitants of Mexico began growing maize, beans, squash, and chili peppers and raising dogs and turkeys. There were no large animals that could be domesticated, however, so all work had to be done by humans. By 2000 BCE, villages dotted the landscape of central Mexico, supporting trade between the different ecological zones.

The most spectacular water control system in the Americas, perhaps in the world, was that found in the Valley of Mexico. There, streams from the surrounding mountains fed a series of shallow lakes. On the edges of these lakes, especially Texcoco and Xochimilco, farmers created *chinampas*, rectangular islands 300 feet long by 15 to 30 feet wide, separated by canals. They did this by dredging up mud from the bottom of the canals and dumping it onto rectangular plots. To keep the soil from washing away, they put up reed barriers and planted willows. Periodically, they added layers of fresh mud and floating vegetation from the canals, thereby keeping the soil fertile. Seeds were sprouted in nurseries and then carefully planted in the chinampas. The abundant fresh water, fertile soil, warm climate, and constant labor allowed the *chinamperos* to grow up to seven crops a year. Each acre of chinampas produced enough food for five or six people, a yield unmatched anywhere else on earth. The earliest chinampas date from the first century BCE, if not earlier. As the population of the valley grew, more and more wetlands were turned into chinampas. In the first eight centuries CE, they supported Teotihuacán, the largest city in the Americas. Even after the fall of Teotihuacán and the rise of the Toltec and Aztec Empires, farmers continued to reclaim land from the lakes.

In the fourteenth century CE, a small tribe called Aztecs took refuge on an island in Lake Texcoco. There, they built the city of Tenochtitlán and proceeded to construct the most elaborate hydraulic engineering project in the Americas. To prevent the salt-laden waters of eastern Lake Texcoco from harming the chinampas to the west of the city during the annual spring floods, they built a ten-mile-long dike across the lake, with gates to control the level of the water. To supply the chinampas and the city with fresh water, they tapped springs in the nearby hills and constructed aqueducts and causeways to the island. Hernán Cortés, who led the Spanish expedition that conquered Mexico in 1519–1521, wrote:

> Along one of the causeways to this great city run two aqueducts made of mortar. Each one is two paces wide and some six feet deep, and along one of them a stream of very good fresh water, as wide as a man's body, flows into the heart of the city and from this they all drink. The other, which is empty, is used when they wish to clean the first channel. When the aqueducts cross the bridges, the water passes along some channels which are as wide as an ox; and so they serve the whole city.[1]

By 1500 CE, on the eve of the Spanish invasion, chinampas covered almost 30,000 acres, providing food for a city of more than 100,000 inhabitants, one of the largest and wealthiest in the world at the time.

In the same period as the rise of civilization in Mexico, another arose along the west coast of South America, where three distinct ecological zones lie in close proximity. The first was the highlands and foothills of the Andes, a region that was cold but received enough rain to

The city of Tenochtitlán, capital of the Aztec Empire, was built on an island in Lake Texcoco. Surrounded by water, Tenochtitlán was so impregnable that the first Spanish attempt to take it ended in failure. In their second attempt, the Spaniards were able to take the city by building boats. Bildarchiv Preussischer Kulturbesitz/Art Resource, NY

grow crops. There, people domesticated llamas, which provided meat and a coarse wool and could be used as pack animals, and alpacas, a smaller species that gave a finer wool. They also cultivated the potato and a grain called quinoa. The second zone was the waters off the coast of Peru. Among the richest fishing grounds in the world, they provided a livelihood to fishermen as far back as 1500 BCE. The third zone was the narrow coastal plain. Although one of the driest regions on earth, it is intersected by rivers that come down from the Andes. Along the banks of the rivers, farmers grew warm-climate crops, such as maize, beans, squash, and cotton. From very early on, the inhabitants of the three zones traded with one another.

Around 1900 BCE, people living along the coastal rivers began digging canals, some of them more than 50 miles long, to bring water and nutrient-rich silt to ever-larger areas of land. Farmers also learned to fertilize their fields with *guano*, the droppings of sea birds that had accumulated for centuries along the coast. In the highlands, farmers built elaborate terraces to grow crops on steep hillsides. The Moche state conquered most of the coastal valleys around 200 BCE and flourished for 800 years, supported by an active trade among the farmers in the rich irrigated lowlands, the herders and farmers of the highlands, and the fishermen along the coast. After 600 CE, the Moche were replaced by two rival civilizations: the Tiwanaku in the southern highlands around Lake Titicaca and the Chimu along the northern coast. By the time the Chimu were overthrown in the 1460s, irrigation canals brought water to millions of acres in more than 60 coastal valleys.

The hydraulic engineering projects of these early civilizations both required and supported large populations. But these civilizations are also known for their building projects and for a rich diversity of crafts that could be produced only by specialists living in settled environments. As Stonehenge and other megaliths attest, the urge to build existed before civilizations arose. But in Neolithic times, such construction took many years because the need to obtain food left the inhabitants with little spare time. In the early civilizations, in contrast, the productivity of agriculture provided a food surplus that could be used to feed construction workers. Furthermore, the habits of cooperation and obedience that came from working together on massive hydraulic engineering projects could be directed by the elites to political and religious construction projects as well.

The earliest building projects undertaken by the Sumerians were temples and cities. They used little wood and no stone but made bricks out of clay and straw and let them dry in the sun. With these sun-dried bricks, they built *ziggurats*, pyramidal towers containing temples, storerooms, and workshops. Baked bricks, too costly for ordinary construction, were used only for decoration. Each temple complex needed professional priests and artisans, merchants, and servants. Cities grew to tens of thousands of inhabitants; the first was Ubaid, built before 4000 BCE.

Land close to a source of water was so valuable that it led to disputes between neighboring cities. As wars broke out, there arose a class of professional warriors supported, like the priests and their retinues, by the surplus from the farms. Wars forced Mesopotamian cities to surround themselves with high walls and gates with heavy doors that could be closed at night or in the event of an attack.

The Egyptians were more fortunate than the peoples of Mesopotamia, for the Nile Valley is bordered by cliffs of good limestone. Stone temples and palaces have survived for thousands of years, whereas ordinary houses, built of sun-dried bricks, quickly melted back into the ground if they were not carefully maintained. The most spectacular constructions in the world, the pyramids of Giza, are almost as good as new after 5,000 years: Khufu, the largest, is 481 feet high and covers

The Sphinx and the great pyramids of Giza are awesome evidence of the ancient Egyptians' mastery of masonry construction. The Sphinx of Giza, carved out of the limestone bedrock, is the largest single-stone statue in the world. Library of Congress LOT 13550, no. 34 [P&P]

13.5 acres; Khafre is almost as huge; and Menkaure is one-third the size of its two great neighbors.

For what purpose were these enormous monuments built? The usual answer is that they were tombs for Pharaohs. Yet one of the earliest Pharaohs in Egyptian history, Sneferu, who reigned from 2613 to 2589 BCE, built three pyramids in succession, two more than he needed as a tomb. The first, at Meidum, began as a step pyramid; an outer mantle, added later to turn it into a true pyramid with 52-degree sides, collapsed into rubble. Next came the Bent Pyramid, so called because it was begun as a true pyramid with 52-degree sides, but once it reached one-third of its intended height, it was quickly finished off at a shallow 43½-degree angle. The third was the Red Pyramid, a true but squat pyramid with 43½-degree sides.

To put huge limestone blocks into place required a labor force of tens of thousands of farmers recruited during the three-month flood season and fed with the grain taken from them as taxes during the previous harvest. As work progressed, however, fewer workers could fit on the top of the growing pyramid. Instead of being dismissed, the others were put to work starting a new pyramid. That is why the Bent Pyramid is bent: it was finished off in a hurry when the architects learned of the collapse of Meidum. Frightened by the disaster, they built the next one, the Red, at a shallow angle. In the process, they mastered the technique of using large stone blocks safely. Only then did they dare to build true pyramids with steep sides, the famous ones at Giza built under Sneferu's successors Khufu and Khafre. In effect, the purpose of pyramid building was to accustom the people of Egypt to cooperate on great construction projects at the behest of their god-king, the Pharaoh. In so doing, Sneferu turned a land of Neolithic farmers into a single nation, Egypt.

The people who irrigated the Indus Valley also built cities. Two of them, Harappa and Mohenjo-Daro, reveal an elaborate but very tightly controlled civilization. Unlike the Mesopotamian cities that grew from villages in a helter-skelter fashion, the two Indus cities were laid out in a rectangular grid, proof that they were planned. They did not have walls but embankments, for they feared not people but floods.

In the Americas, long before cities appeared, civilization was associated with the building of large ceremonial centers where few people lived year round but to which many came on special holidays. In the first millennium BCE, the Olmecs of Mexico carved gigantic stone statues weighing up to 20 tons and transported up to 100 miles from where they were quarried. By the first century BCE, the temples and pyramids of Monte Albán, in the Valley of Oaxaca, attracted enough merchants, artisans, and other nonfarmers to qualify as a town. Likewise, the Mayans of southern Mexico and Guatemala created temple complexes such as Tikal surrounded by villages with several thousand inhabitants.

The first true city in the Americas was Teotihuacán in the Valley of Mexico. Founded around 200 CE, it flourished from 300 to 700 but then declined. The people of the region built two great pyramids, the Temple of the Sun and the Temple of the Moon, along with hundreds of smaller pyramids, temples, and religious or political buildings. Around them, they laid out a city in a rectangular grid, with neighborhoods devoted to artisans in obsidian, pottery, cloth, leather, and bird feathers and inhabited by merchants from other parts of Mexico. In its heyday, Teotihuacán had close to 100,000 inhabitants.

In other parts of the Americas, as in Mexico, ceremonial centers preceded cities. El Paraíso in Peru, built about 1800 BCE, included six huge buildings and required 100,000 tons of stone. Not until 2,000 years later was the first true city, Chan Chan, built in South America. In the southwestern part of the United States, the Ancestral Pueblo (or Anasazi) people built several ceremonial centers such as Pueblo Bonito in Chaco Canyon, New Mexico, or the more famous Cliff Palace in Mesa Verde, Arizona, with its 220 rooms and 23 *kivas,* or circular religious centers. These centers had only a small permanent population but served as meeting places on special occasions for thousands of people from outlying villages.

Not all the technologies of the early civilizations were as grandiose or required as much cooperative effort as water control systems or cities and monumental buildings. Some were on a smaller scale, yet were just as important to the lives of the people. Two of these, weaving and pottery, were useful to everyone, even the poorest. Others, like metallurgy and wheeled vehicles, were mainly of interest to the upper classes.

Unlike hunters and gatherers who clothed themselves in animal skins, agricultural people needed textiles. In every civilization, weaving cloth was done by both men and women, but spinning yarn was always the work of women. In the Hebrew Bible, the virtuous woman "seeketh wool and flax and worketh willingly with her hands. She layeth her hands to the spindle, and her hands hold the distaff" (Proverbs 31:13, 19, 24). The distaff was a long stick that held the roving, or loose fibers, while the spindle was a short stick that rotated as it dropped, giving the yarn a twist as it wound it. Using these simple devices, women could spin yarn while walking or carrying out other tasks. To this day, the words *distaff* and *spinster* reflect this ancient women's occupation.

The string skirts and belts made by Neolithic peoples were of silk, hemp, and flax. By the fourth millennium BCE, the people of Mesopotamia wove woolen cloth from the inner fleece of goats and of sheep; later, they bred sheep specifically for their fleece. Wool was first made by pounding the fibers together into felt. Documents written in the Sumerian city of Ur describe flocks of sheep and the female occupations of spinning and weaving. Around 2300 BCE, herdsmen on the Mediterranean island of Crete began breeding sheep with long woolly fleece that could be combed, spun into yarn, and woven into cloth. Unlike hemp and flax, wool yarn came naturally in a variety of colors ranging from black to white and moreover could be dyed in vivid colors. The women of Crete and nearby Greece began producing colorful textiles, including new weaves such as twill and tapestry, that were traded throughout the eastern Mediterranean. The Greek poet Homer wrote: "And fifty women serve Alcínoüs: some grind the yellow corn, their mill in hand; and others weave their webs or, while they sit, twist their swift spindles as their fingers glint like leaves of poplar trees swayed by the wind. . . . As the Phaeácians are the most expert of men in sailing brisk ships over seas, so are their women peerless when they weave."[2]

Wool was also the principal textile used by the people of the highlands of Peru. Ordinary people wore cloth made from the wool of llamas. The cloth worn by the elites came from the wool of alpacas raised specifically for that purpose; it was the finest and most skillfully woven cloth in the world, with up to 500 threads per inch, and came in many colors with elaborate designs of birds, jaguars, and snakes.

Egyptians, alone among the peoples of antiquity, dressed almost exclusively in linen garments made from flax. Fragments of Egyptian linen date back to 4500 BCE. Egyptians used linen not only to clothe the living but also to wrap mummies for burial and to make sails for the boats that plied the Nile. Men grew, harvested, and prepared the flax fibers, and both women and men wove the cloth. Much weaving was done in large workshops where workers toiled under the watchful eye of male—and sometimes female—overseers. The Greek historian Herodotus was shocked: "Their habits and their customs are the exact opposite of other folks'. Among them the women run the markets and shops, while the men, indoors, weave. . . ."[3] Since linen could not be dyed, wealthy Egyptians wore white clothes decorated with colorful jewelry made of gold from sub-Saharan Africa, silver from the Aegean, lapis lazuli from Persia, amber from the Baltic, and other precious materials carried over long distances by merchants.

In China as in western Eurasia and North Africa, women also wove cloth; before the fourth millennium BCE, it was made from hemp or ramie, both vegetable fibers. Then the Chinese learned to raise the silkworm *Bombyx mori* that eats only mulberry leaves. Making silk involves boiling the cocoons, reeling or unwinding the fibers, and twisting them into threads, all delicate hand operations. The result is a wonderfully strong and smooth fiber that can be dyed in brilliant colors and woven into many patterns such as tabby and brocade. Although farmers most probably discovered this fiber, Chinese legend attributes it to the mythical Empress Xi Ling-shi. As in other cultures, the Chinese divided tasks by gender; as their saying had it, "men till, women weave."[4] By the Shang dynasty, aristocrats dressed in silk robes while ordinary people clothed themselves in hemp, ramie, or cotton. For several thousand years, the secret of raising silkworms and making silk was known only to the people of China.

Cotton has two origins: India and the Americas. The people of the Indus Valley cultivated cotton and wove cloth as early as 3500 BCE. In 700 BCE, travelers brought the plant to Mesopotamia, from where its use spread to Egypt and sub-Saharan Africa. Another variety of cotton was used by the peoples of the lowlands of Peru and of Mesoamerica, tropical regions where the plant grew well and the cloth was better suited to the climate than wool.

Next to textiles, pottery was the most important craft of ancient times. The history of pottery has been carefully studied, for pottery shards, unlike textiles, do not deteriorate when left in the ground. Because pots are heavy and fragile, they were of no use to nomadic peoples. The very first potters, the Jomon people of Japan, were unique among foragers in having permanent villages. Elsewhere, pottery was a characteristic of Neolithic peoples and of civilizations.

The earliest potters after the Jomon were the inhabitants of the Zagros Mountains, north of Mesopotamia. Beginning in the seventh millennium BCE, they made pots by forming spirals of clay mixed with straw or grit that they then smoothed out, dried, and fired in a bonfire or a kiln. Pottery had many uses. Large pots were used to store grain, oil, and wine. People used smaller ones to carry water, to cook in, and to drink and eat out of. Like spinning and weaving, pottery making was a domestic art that created domestic objects; in many cultures, it was practiced by women. Pots were often decorated, each culture and period creating its own distinctive pattern. This has allowed archaeologists to identify the origin of pieces of pottery found in the ground and to trace the diffusion of styles and the patterns of trade; pieces of Indus

Valley pottery have been found in Mesopotamia, and Chinese pottery was traded throughout East Asia.

Metallurgy began with the use of native copper and gold. Such were the pieces of copper that the inhabitants of the Great Lakes region of North America and of Anatolia (now Turkey) hammered into knives, chisels, axe blades, and arrowheads. The axe blade belonging to the Ice Man of the Alps was of native copper. Native metals are rare, however. In nature, most metals exist in the form of ores or minerals containing metal oxides and other compounds. Stone Age foragers were familiar with many different kinds of stones such as malachite and azurite that ooze metallic copper when heated. Traders deliberately brought such ores to Mesopotamia by the fifth millennium BCE and later to Egypt, the Indus Valley, and northern China. Though fairly common, copper ores are concentrated in certain locations such as Anatolia, Iran, the Sinai Peninsula, Oman, Cyprus, and Nubia. Thus, ores and metals were among the most important items of long-distance trade.

Mining ores and turning them into copper were complicated processes. Surface deposits were quickly exhausted, and underground ores tended to be sulfides (compounds containing sulfur) that had to be crushed and roasted before they could be smelted. Smelting required furnaces with a strong draft and a great deal of fuel. The earliest furnaces were not hot enough to melt the metal, which had to be repeatedly heated and hammered into the required shape. Bellows, to create a blast of air and increase the temperature of the fire to the melting point of copper (1,200 degrees Celsius), did not come into use until the late second century BCE. After that, it became possible to cast the molten metal into molds of the desired shape.

The many skills involved in making metal objects turned their practitioners into full-time specialists—miners, smelters, smiths, and others—who jealously guarded their secrets and passed them on from father to son. They made axe and adze heads, saw blades, drainpipes, knives, swords, armor, and many other objects. Such items were very costly, however, and reserved for the elites. Farmers and ordinary artisans continued to use tools of stone and wood long after copper became available to the wealthy.

Although copper is easy to work, it tends to become hard and brittle with use and needs to be reheated and reshaped periodically. In the third millennium BCE, metallurgists learned to mix copper with softer metals to create alloys that were easier to work yet stronger and more durable than pure copper. The first of these was arsenic, often found in nature mixed with copper ores, producing an alloy called arsenical bronze. Arsenic is a poison, however, and it was soon replaced by tin, which was less dangerous to work with. Bronze made of copper and tin (usually in a 10:1 ratio) was a superior metal in every way; it had a lower melting point than copper, was easy to cast, did not become brittle, and kept an edge. Tin ores, however, were rare and had to be imported to the Middle East from as far away as England. Although their names have long since been forgotten, we know that merchants traded over long distances because archaeologists have found Middle Eastern pottery shards thousands of miles from where they were manufactured.

Bronze was well known in the Middle East, North Africa, and Europe, but it was Chinese metallurgists who perfected the lost-wax method of casting the metal into intricate shapes. They first created a rough model of the desired object in clay. Once it was dry, they covered it with wax into which they carved fine details. They then coated the wax carving with clay and again allowed it to dry, forming a mold. They then heated it, allowing the molten wax to run out through holes left in the mold. This left a void inside the mold into which they poured molten bronze that took on the exact shape vacated by the wax. When the mold cooled, they cracked it open, revealing inside an exact replica in bronze of the wax carving. The aristocrats of the Shang dynasty period controlled access to the copper and tin mines and managed the smelters and workshops in which bronze was worked. Under their direction, bronze smiths made intricately decorated ritual vessels, musical instruments, chariot fittings, and especially the weapons with which the nobles maintained their power over the rest of the population.

The Americas also had centers of metallurgy. Native copper, gold, and silver were found in several places in the New World. Peruvian smiths fashioned gold into decorative items as early as 500 BCE. By the first century BCE, smiths had learned to cast copper in Colombia and later in Peru and Mexico. By the fifth century CE, smiths were casting objects of bronze. On the eve of the Spanish invasion, metalsmiths were using the lost-wax method and making beautiful objects of gold, silver, and platinum, as well as knives, weapons, and tools of copper and bronze.

The inhabitants of these ancient civilizations were remarkably creative in many fields, but one technological artifact they were slow to develop was the wheel. In today's world, the wheel is so ubiquitous and indispensable that it is hard to imagine a world without it. Yet even when the people of early civilizations understood the principle of the wheel, they made little use of it.

The Shang culture made tremendous progress in bronze working and other military technologies. This object may have been a chariot fitting. Gift of Charles Lang Freer, F1911.89, Freer Sackler Gallery

The potter's wheel made its appearance in Mesopotamia in the fifth millennium and was adopted in many parts of Eurasia. Around 3700 BCE, cattle herders who lived on the plains north of the Caucasus Mountains began burying their leaders with two-wheeled carts or four-wheeled wagons drawn by oxen, signs of wealth and power rather than utilitarian vehicles. Their wheels, made of three heavy planks encircled by leather straps with big copper nails, were firmly attached to their axles, which turned with them. Such vehicles became common in Mesopotamia and Syria by 3000 BCE and in the Indus Valley 500 years later. They were also known in northern Europe by 3000 BCE and in Egypt after 1650 BCE. However, they were very heavy and bogged down in soft soils and could not be used on rocky terrain. Long after the wheel was known, it was much easier to transport goods over long distances in caravans of donkeys. The peoples of the Americas did not use wheeled vehicles at all because they had no domesticated animals large enough to pull them.

One of the most important technologies we have inherited from the ancient civilizations is writing, a means of storing and transmitting information through space and time by inscribing symbols to represent things, ideas, and sounds. Many different writing systems have appeared in the world. Those of Mesopotamia, China, and Mesoamerica were independently invented. Others, like the Egyptian hieroglyphs and our own alphabet, were inspired by the writings of neighboring societies.

The Sumerians created the first writing system, called *cuneiform*, meaning "wedge shaped." Around 8000 BCE, people in and near Mesopotamia began using small clay tokens to represent such things as sheep, bushels of grain, or jars of oil. Meanwhile, others were drawing designs on pottery. Between 3300 and 3200 BCE, Sumerian scribes began depicting not only people and things but also abstract ideas. They used the rebus principle, like drawing a picture of a bee and a leaf to indicate the word *belief*. They inscribed these symbols with a stick with a wedge-shaped end on small tablets of wet clay. Once dried in the sun, these tablets lasted for thousands of years.

For the first 500 years after cuneiform was developed, it was used only to make lists, keep track of donations to temples, and the like; 90 percent of the tablets found are bookkeeping and administrative documents. Only later did scribes begin writing histories, laws, legends, and other forms of literature. To do so, they needed 500 to 600 different signs, requiring many years to learn. Only a very few people had the leisure to learn this esoteric skill or the wealth to send their

A cuneiform tablet is enclosed in a clay envelope with an inscription and seal, dating from about 2000 BCE and found in Turkey. Consisting of a combination of wedge shapes, cuneiform permitted a high level of commercial and government transactions. The figures at the top, impressed into the moist clay with a cylinder seal, represent a man standing (probably a king) presenting a gift or tribute to a larger seated figure, who is most likely a god. The inscription is written vertically from top to bottom. Library of Congress LC-USZ62-82973

children to school. Writing became a way to distinguish the literate elite from the rest of the population.

The early civilizations brought forth many admirable innovations not only in agriculture and construction but also in all the arts, crafts, and sciences. These civilizations are traditionally extolled as "the dawn of history." There are good reasons to celebrate their accomplishments, for our own civilization is based on theirs.

But the evidence is loaded, as is the very word *civilization*. All the writings and monuments and most of the artifacts left by these ancient civilizations were created by or for their elites. The historians who have described them, and the readers of their works, are also members of literate elites. For them and for us, civilization represents a great advance over the lives of the hunter-gatherers and Neolithic farmers who preceded them.

Why, if civilization represented such an advance, did it remain restricted to a few regions and not spread to the rest of the world for many centuries? This question is usually asked about the indigenous peoples of North America, Africa, and Australia. The same question can be asked about the Greeks and other Europeans, who resisted the attractions of civilization for 2,000 years after the beginnings of Egyptian civilization. Were they somehow "retarded"?

In fact, quite the contrary is true. The ancestors of today's Europeans were Neolithic farmers who lived in Anatolia some 10,000 years ago. As their food supply increased, so did their numbers, and some of them began to migrate, seeking fertile rain-watered lands. Wherever they settled, their numbers soon exceeded those of the indigenous foragers. Long after civilizations had emerged in Mesopotamia, Egypt, and the Indus Valley, these early farmers were still finding new lands to farm. They deliberately avoided becoming civilized as long as possible. Not until the first millennium BCE did the inhabitants of Greece and Italy occupy all the arable land and find themselves in the same predicament that had led the peoples of the Middle East to adopt civilization. Only then did they submit to the discipline of taxes, laws, and religious or political authorities.

Thus, the technologies that characterize the early civilizations represented a great advance in the power of human beings over nature but also in the ability of a small elite to impose their rule on large numbers of their fellow humans.

CHAPTER 3

Iron, Horses, and Empires (1500 BCE–500 CE)

Pliny the Elder, a Roman nobleman who lived from 23 to 79 CE, at the height of his country's power, looked back on the invention of iron with great misgivings. What he wrote about iron could also be said of many other technologies.

> We will now consider iron, the most precious and at the same time the worst metal for mankind. By its help we cleave the earth, establish tree-nurseries, fell trees, remove the useless parts from vines and force them to rejuvenate annually, build houses, hew stone and so forth. But this metal serves also for war, murder and robbery; and not only at close quarters, man to man, but also by projection and flight; for it can be hurled either by ballistic machines, or by the strength of human arms or even in the form of arrows. And this I hold to be the most blameworthy product of the human mind.[1]

The iron to which Pliny refers—and another innovation, the domestication of horses—disrupted the old river valley civilizations of Eurasia. These civilizations had lasted for more than 1,000 years with only minor changes because their religious and political elites had little to gain and much to lose from innovations that could threaten their status. Yet they had no control over the rain-watered lands outside their borders. It is in these marginal areas that innovations appeared. At first, iron and horses allowed people who possessed them to conquer or disrupt the older civilizations. Eventually, the nomadic warriors and the settled river-valley peoples merged to form new empires, vast areas with millions of inhabitants ruled by centralized governments. Such were the empires of Assyria, Persia, Rome, China, and India.

To hold these vast empires together and to encourage their inhabitants to accept distant, often foreign, rulers required not just powerful armies and elaborate bureaucracies but also new technologies. Imperial governments devoted resources to great public works projects, such as roads, aqueducts, and walls. They also encouraged shipbuilding and shipping and, to a lesser extent, crafts and mechanical engineering.

The first people to smelt iron, the Hittites, lived in Anatolia some time after 1500 BCE. Smelting iron ore was much more difficult than smelting copper or bronze. The simple furnaces used at the time—pits dug in a hillside and lined with stones or clay—could not get hot enough to melt iron. What came out was a spongy mixture of iron and slag (or dirt) known as a *bloom*. To drive out the slag, blacksmiths had to heat and hammer the bloom repeatedly, a tedious process that required a great deal of time and charcoal. The result—wrought iron—was softer than bronze, cracked easily, did not hold an edge as well, and rusted rapidly. Yet iron had one tremendous advantage: its ores are found in large quantities in almost every country, often close to the surface, where they are easy to dig up.

Gradually, by trial and error, blacksmiths improved their product. To prevent cracking, they learned to cool hot iron by dipping it in cold water. To make it less brittle, they tempered it by reheating the quenched iron several times. By repeatedly placing an iron object in direct contact with burning charcoal, they turned its surface into steel. By the fourth century BCE, blacksmiths were making iron swords with a steel cutting edge that was hard enough to cut through bronze.

Iron transformed both war and peace. For a few centuries, it gave the Hittites an advantage over their enemies. When the Hittite Empire collapsed in 1200 BCE, the technology spread to other Middle Eastern countries. The disruptions that followed hampered the imports of tin from Britain, thereby hastening the switch to iron.

Once blacksmiths had mastered the art of smelting iron ore, their products became much cheaper than copper or bronze products. Even peasants and artisans could afford metal tools. Blacksmiths used iron tongs and anvils; farmers used iron plowshares, axe heads, and hoe blades; carpenters had iron saws, chisels, files, and adzes; cooks used iron knives and pots and pans. Even lamps and furniture were made of iron. For most people, it was iron, not bronze, that brought an end to the Stone Age. Not only did iron smelting become common in established centers of civilization, it also allowed farmers outside these areas to clear forests with iron axes, spreading agriculture to new parts of the world.

Iron making reached India from Mesopotamia soon after 1000 BCE. Armed with iron axes, farmers felled the thick forests that covered the Ganges Valley and then plowed the land with oxen. Gradually, farming spread to the rest of the subcontinent once covered by forests. The Ganges Valley, warm and wet, lent itself to wet-rice agriculture, with much higher yields than the wheat and barley that grew in the Indus Valley. Soon the area was covered with paddies and permanent villages, homes to a dense population.

Knowledge of iron smelting reached China around 700 BCE, and iron was in widespread use two centuries later. Chinese foundries carried the technology of iron smelting far beyond any other in the world. Iron masters built large furnaces equipped with bellows to force air through the burning mixture of ore, charcoal, and lime. Later, they attached the bellows to waterwheels turned by flowing streams. By the fourth century BCE, these improvements raised the temperature of the furnace to the melting point of iron (1,537 degrees Celsius). Molten iron could be cast in molds to make pots and pans, bells, sickles and statues, and many other objects. Cast iron, though not flexible, was much harder than wrought iron. The Chinese were casting iron 1,000 years before any other people. They also found ways to heat the iron with coal, a technique not reproduced elsewhere for 2,000 years. Farmers with iron axes, hoes, and plowshares brought irrigated rice agriculture to the Yangzi Valley in central China. Soon, the population of central and southern China rivaled that of the Yellow River Valley, where Chinese civilization began.

Iron reached Africa from three directions. The Egyptians learned iron smelting from the Assyrians, a Mesopotamian people who conquered Egypt in the seventh century BCE. From there, the craft spread to Kush, south of Egypt. For several centuries, Meroë, the capital of Kush, manufactured iron goods, leaving huge slag heaps behind. Blacksmiths from Arabia brought iron making to Ethiopia in the ninth or tenth century BCE. A third path was from Palestine to Carthage in Tunisia; from there, travelers crossing the Sahara introduced iron-making skills to northern Nigeria in the first century BCE.

Until then, Africa south of Ethiopia and northern Nigeria were sparsely occupied by hunter-gatherers. Copper was known but little used. It was iron that transformed the continent. The first Africans south of the Sahara to smelt iron were inhabitants of northern Nigeria and Cameroon who spoke languages of the Bantu family. The smelting of iron was done by men. It was often restricted to certain families or clans and involved secret rituals. Women gathered, sorted, and washed the ores, prepared charcoal, and made food and beer for the iron workers but were not allowed to smelt or forge iron. Pottery making, a technology similar to iron smelting—taking raw material from the ground, preparing it, and baking it in a fire—was almost everywhere the work of women, often the wives of iron workers.

With iron, Bantu-speaking Africans were able to fell trees and open up forested areas to cultivation. They used iron for the blades of hoes, much more efficient tools to till the soil than wooden digging sticks. Felling trees was most often men's work, while cultivating crops was more often done by women. Iron-using farmers grew crops specific to the tropical climate: sorghum and millet in dry areas, rice and yams and later bananas in rainy regions. They built permanent villages and made cloth and pottery. In grassland areas too dry to farm, nomadic herdsmen raised cattle. Using these new technologies and skills, Bantu-speaking agriculturalists migrated outward from their original homeland. By the end of the first millennium CE, the majority of the peoples in eastern, central, and southern Africa spoke Bantu languages. Displaced by the newcomers, foragers retreated into ever more remote forests and deserts, where a few still survive by hunting and gathering.

In the Sudanic grasslands where horses thrived, the combination of horses and iron weapons led, as in Eurasia, to the rise of kingdoms. When the first Muslim traders crossed the Sahara Desert in 753, they found well-established states dominated by kings and an aristocracy of mounted warriors. The Arab traveler al-Muhallabi, who visited the central Sudan in the tenth century, disapproved of the adulation Africans showed toward their king: "Him they respect and worship to the neglect of Allah the most high. . . . Their religion is the worship of their kings, for they believe that it is they who bring life and death and sickness and health."[2]

Wild horses, small, tough animals native to the grasslands of central Asia and southern Russia, were domesticated in the third millennium BCE by nomadic peoples. They were poorly suited for pulling a plow or a heavy cart, however, because the throat-and-girth harness used in western Eurasia choked them. The peoples of Ethiopia, North Africa, and parts of West Africa also raised horses, but horses could not survive elsewhere in Africa because of tsetse flies that carried the deadly sleeping sickness. Sometime in the mid-second millennium, carpenters in the Caucasus began making wheels with four or six spokes that were much lighter and less fragile than solid wheels. Such wheels permitted them to build lightweight two-wheeled chariots pulled by two horses and carrying both a driver and an archer. This invention inaugurated a period of warfare in which fast and deadly chariots decimated defenseless foot soldiers. Between 1700 and 1300 BCE, charioteers from southern Russia and central Europe conquered older civilizations and established short-lived kingdoms that engaged in frequent warfare; the Hyksos who invaded Egypt and the Aryans who invaded northern India were among such chariot-mounted warriors.

In warfare, revolutionary innovations in weaponry are never permanent, for they stimulate the development of counterweapons. Chariots, which appeared when bronze was the predominant metal, were first challenged, and then made obsolete, by a combination of iron and cavalry.

Around the end of the second millennium BCE, nomads from southern Russia bred horses large and strong enough to carry a man. They also learned how to control a horse with their legs while shooting arrows or wielding a lance or a sword. Like chariots before them, cavalry proved dangerous to all but the best armed and organized infantrymen. The possession of horses and cheap iron weapons encouraged warfare throughout Eurasia. Cavalrymen, many of them nomads, saw opportunities to attack and invade settled agricultural regions. Nomads the Egyptians called Sea People invaded Egypt, and others called Dorians destroyed the first Greek civilization in Crete. Nomadic horseback-riding Phrygians from southern Russia overthrew the once powerful chariot-riding Hittites. In Mesopotamia, the rising Assyrian Empire armed its infantrymen with iron swords and battering rams. With these new weapons, its peasant soldiers conquered Mesopotamia, Syria, Palestine, and Egypt between the tenth and seventh centuries BCE.

After about 900 BCE, India was divided into small kingdoms dominated by the descendants of Aryan warriors who had invaded the subcontinent around 1500 BCE. While the peasants toiled away in their fields and paddies, members of the warrior caste armed with iron weapons fought one another for control of the land. Finally, in the fourth century BCE, Chandragupta Maurya conquered an empire that stretched from the Indus to the Bay of Bengal and encompassed most of the subcontinent. Although the dynasty he founded lasted little more than a century, it was the largest kingdom in Indian history before the nineteenth century.

Iron proved equally disruptive in China. The Zhou (pronounced "joe") dynasty that had succeeded the Shang in 1122 BCE disintegrated into rival provinces led by local warlords. In a period called the Warring States (403–221 BCE), massive infantry armies battled with iron weapons and armor. Finally, in 221 BCE, Shi Huangdi (pronounced "she wongdee"), the leader of Qin ("chin"), the most aggressive of the warring states, conquered all the others. Although it hardly survived its founder, the Qin dynasty—from which we get the word China—is

An army led by the ninth-century BCE *Assyrian King Ashurnazirpal batters a Mesopotamian city with siege machines. Siege machines were designed to allow soldiers to come close enough to a city to ram the walls or gates with a heavy beam, while protecting them from defenders' arrows. Such machines were often as tall as the city's walls.* Erich Lessing/Art Resource, NY

remembered as the most despotic government in Chinese history but also the one that unified China.

It was the next dynasty, the Han (202 BCE to 220 CE), that reaped the benefit of iron and national unity. For four centuries, Han emperors ruled over a large and prosperous land. They did not leave iron making to individual blacksmiths but established large foundries with thousands of workers. Casting molten iron into durable molds, these foundries turned out identical objects in large quantities, the first known instance of mass production.

In western Eurasia, the Romans used more bronze than iron. Their infantrymen, organized in legions of 4,200 to 6,200 men, wore armor of bronze or iron plates fastened with leather and carried iron swords and javelins, as well as tools with which to build palisades and dig ditches around their camps. On the eastern frontier of their empire, however, the Parthians, a nomadic people from southern Russia whose cavalrymen and horses were fully armored, defeated the Romans, while Germanic tribes repeatedly invaded the northern frontier of the empire.

Despite the success of their empires, their technological underpinnings—iron and horses—could not be monopolized by states or civilizations. Instead, these technologies gave nomadic herdsmen and warriors the weapons to harass and sometimes defeat the civilized empires. The conflicts between settled empires and nomadic herdsmen are one of the themes of Eurasian history for more than 1,000 years.

From 1500 BCE to 500 CE, four major regions of Eurasia—southern Europe, the Middle East, India, and China—were ruled by successions of empires, each one larger than its predecessor. Population growth and new technologies, especially iron, made agriculture in rain-watered areas productive enough to support cities, governments, and armies—in other words, civilizations. To gather tribute from distant provinces and maintain their control over them, imperial governments needed efficient means of communication and transportation. With the wealth collected from all their provinces, they built roads, cities, and elaborate public works. This was a great age of construction and civil engineering.

In civil engineering, the champions of that era were the Romans, masters of the Mediterranean world from the third century BCE to the fifth century CE. Not only did they devote more resources to buildings and public works than any culture since early pharaonic Egypt, but they also introduced several important innovations. One was the arch, made of wedge-shaped stones held up on either side by pillars. The Romans prolonged the arches to form barrel vaults, allowing them to build halls up to ninety feet wide without interior columns. Having mastered the vault, they learned to make two vaults intersect in the form of a cross and then a dome or half-sphere resting on thick walls. Two of their most impressive structures were the Pantheon, a domed building with an interior diameter of 142 feet built between 110 and 125 CE, and the Colosseum, a stadium 600 feet long by 175 feet high able to hold 50,000 spectators, which was built from 72 to 80 CE.

One of the reasons the Romans were able to make large buildings and other structures was their discovery of *pozzolana*, a volcanic earth that when mixed with lime formed a cement that was not affected by water or fire. The technique of making cement was forgotten after the fall of Rome in the fifth century and not rediscovered until the nineteenth century. Many of their concrete structures, often faced with brick or stone, are still standing 2,000 years later.

To supply their cities with fresh water and to remove polluted water and sewage, Roman engineers created the most elaborate water-supply systems in the ancient world. They studied the works of the Greeks, who built short aqueducts near their towns. Petra, in

Jordan, was another role model; though located in the middle of a desert with an average annual rainfall of less than four inches, this city of 30,000 inhabitants was graced with baths, pools, and gardens watered by an elaborate system of catchment basins and ceramic pipes.

The Romans built water-supply systems on a vastly larger scale than their predecessors. To supply the city of Rome, they built nine canals with a total length of 300 miles bringing fresh water from streams in the Apennine Mountains. These canals wound their way along hillsides, following a steady gradient of two to three inches to the mile. To cross valleys, the Romans built aqueducts, many of which are still standing. The Pont du Gard, in southern France, is 900 feet long and 160 feet high at its peak; the aqueduct of Segovia, in Spain, is half a mile long and still in use. In places where a valley was too deep to bridge, they built inverted siphons, pipelines made of lead running down one side and up the other. Once the water reached its destination, it was stored in reservoirs and piped to fountains, public baths, and the homes of the

The ruins of an aqueduct in Constantine, Algeria, testify to the hydraulic expertise of the Romans, who constructed them throughout their expanding empire. Library of Congress Lot 13553, no. 12 [P&P]

rich. This permitted most city dwellers to bathe daily, a practice that was forgotten until the twentieth century.

The Romans were rightly proud of their achievements. As Sextus Julius Frontinus, water commissioner of Rome from 97 to 103 CE and author of the engineering manual *De aquis orbs Romae*, exclaimed: "With such an array of indispensable structures carrying so many waters compare, if you will, the idle Pyramids or the useless, though famous, works of the Greeks."[3]

The hydraulic engineering of the Chinese was as impressive as that of their Roman contemporaries, but its purpose was to supply water for agriculture and navigation rather than for cities. In northern China, hydraulic engineers concentrated on flood control. In the tropical lowlands of central and southern China, rice yielded much more food per acre than any other crop but only if it was irrigated. The efforts of people and governments focused on building dams, canals, and terraces to grow rice wherever possible, even on steep hillsides.

The first major irrigation project was built in Sichuan, in central China, in the third century BCE. But it was the Qin Emperor Shi Huangdi who inaugurated some of the great construction projects for which China is famous. One of his projects, the 20-mile-long Magic Canal, is still in use after 2,200 years. To permit the efficient transport of rice and other products from central China to the north, where the government was located, he began construction of the Grand Canal, which later emperors extended to link the Yellow and Yangzi River Valleys. At the same time, Shi Huangdi began building a wall of hard-packed earth to protect the settled agricultural areas of northern China from the nomadic herdsmen who lived further north, in the grasslands of Mongolia, and periodically attacked the farming communities. This was the origin of the famous Great Wall that later emperors completed over the next 2,000 years.

Not all construction projects were hydraulic. To remain unified, all empires needed good transportation. Road building was therefore an essential activity, partly for trade, but more important, to allow rapid communication and troop movements between the capital and outlying provinces. Not surprisingly, the Assyrians, who dominated the Middle East from the tenth to the sixth centuries BCE, built a road network to hold their empire together soon after they began to use horses for long-distance communication. The Persians, who succeeded them as the dominant empire in the Middle East, completed the Assyrian road network between the sixth and the fourth centuries. Their Royal Road stretched 1,677 miles from Susa, in southern Iran, to Ephesus on the

Aegean Sea. It was maintained and patrolled by soldiers. While merchant caravans and ordinary travelers took three months to cover that distance, royal couriers could make the journey in one week. The Greek historian Herodotus was impressed:

> Nothing mortal travels so fast as these Persian messengers. The entire plan is a Persian invention; and this is the method of it. Along the whole line of road there are men (they say) stationed with horses, in number equal to the number of days which the journey takes, allowing a man and a horse to each day; and these men will not be hindered from accomplishing at their best speed the distance which they have to go, either by snow, or rain, or heat, or by the darkness of night. The first rider delivers his despatch to the second, and the second passes it to the third; and so it is borne from hand to hand along the whole line, like the light in the torch-race.[4]

Other empires built roads for the same reasons the Persians did. In China, Shi Huangdi ordered roads to be built with special lanes reserved for official couriers. He also standardized the width of axles of all carts and chariots to fit the ruts in the soft earth. The Han emperors extended his road network and paved it with gravel. In the first century BCE, traders opened a route between Changan, the Han capital, and the oases of central Asia. This route, which we call the Silk Road, was equipped with relay stations for official couriers and guest houses for travelers.

The Romans built the most extensive network of roads in Eurasia. Their first road was the Via Appia between Rome and Capua, near Naples, built in 312 BCE. The next one, the Via Flaminia, went north from Rome to Rimini. In later centuries, they built 50,000 miles of roads throughout their empire. Road builders were highly regarded in Roman society; Vitruvius, engineer, architect, and author of the classic work *De Architectura*, described an engineer as "a man of letters, a skillful draftsman, mathematician, one familiar with history, philosophy, music (so as to know how to tune catapults by striking the tensions skeins) . . . not entirely ignorant of medicine, familiar with the stars, astronomy, calculations. . . ."[5]

Roman roads were four feet thick and cambered so water would drain off. They consisted of five layers: a bottom layer of sand, then one of flat stones, then gravel set in clay or concrete, then more concrete, and finally large cobblestones set in concrete. The roads were as straight as possible to minimize the distance their heavily armed legionnaires had to march. Bridges rested on concrete piers spaced up to 100 feet apart with semicircular arches between them. Along their highways ran the official postal service, the *cursus publicus*, with relay stations every 10 to 12 miles and inns for travelers every 30 to 40 miles. Merchants also used the roads to transport goods in carts or on the backs of horses, mules, and donkeys. Like their aqueducts, the Roman roads and bridges were so solidly built that many are still in use today.

Until the mid-nineteenth century, transportation on water was easier, less expensive, and safer than on land. We do not know what the first boats were like; possibly, they were carved out of logs or made of several logs tied together into a raft. The first craft that we can document were boats of papyrus reeds in Neolithic Egypt. In pharaonic times, the Egyptians made wooden boats with simple square sails to carry them up the Nile when the north wind blew. To go north, they furled the sails and let the current carry them. Some were large enough to carry stone blocks for the pyramids. In Mesopotamia, people built floats of animal skins that were easy to carry back upriver.

By the second millennium BCE, small boats were sailing the Mediterranean, the Red Sea, and the Persian Gulf. Some sailors ventured as far as India, hugging the coast and beaching their boats at night. Gradually, they took more risks. In the first millennium BCE, Greeks and Phoenicians—a people who lived along the coast of Palestine—began navigating the Mediterranean on a regular basis. Because the soil of their homeland was too thin for wheat, the Greeks grew grapes and olives and traded wine and oil for wheat from Sicily and the shores of the Black Sea. To do so, they became sailors. Later, the Romans seized the Mediterranean Sea from both the Carthaginians, heirs of the Phoenicians, and from the Greeks. They called it *Mare Nostrum*, "our sea."

The seafarers of the Mediterranean—Greeks, Phoenicians, and Romans—built both merchant ships and warships. Their merchantmen were broad bottomed and partly decked, with a square sail on a single mast and two rudders, one on each side. They sailed well before the wind but had great difficulties tacking upwind. A Roman cargo ship could sail from Sicily to Egypt in a week if the wind was behind it but took 40 to 70 days to make the return voyage. So ill suited were these ships to sailing in unfavorable winds that all navigation ceased from October through April, when the sea was stormy and the winds uncertain.

In contrast to the slow and heavy cargo ships, warships were light, fast, and propelled by rowers. Early Greek warships, called *pentekonters*, carried 50 oarsmen, each pulling an oar; they could reach a top speed of 11 miles per hour, similar to that of a modern racing shell. By the sixth century BCE, the Athenian navy was equipped with *triremes*, long narrow ships with three rows of oars on each side (hence the name) pulled by 170 rowers, closely packed together and trained to row in perfect unison. For brief moments, triremes could reach a top speed of 13 miles per hour. They carried bronze rams in front to puncture the hulls of enemy ships. Naval battles involved ramming, boarding, and hand-to-hand combat. This is how the Greek historian Thucydides described a key moment in the battle between the navies of Athens and Syracuse in 413 BCE in *The Peloponnesian War:*

> So long as a vessel was coming up to the charge the men on the decks rained darts and arrows and stones upon her; but once alongside, the heavy infantry tried to board each other's vessel, fighting hand to hand. In many quarters it happened, by reason of the narrow room, that a vessel was charging an enemy on one side and being charged herself on another, and that two or sometimes more ships had perforce got entangled round one, obliging the helmsmen to defence here, offence there, not to one thing at once, but to many on all sides; while the huge din caused by the number of ships crashing together not only spread terror, but made the orders of the boatswains inaudible.[6]

With three rows of oars on each side, triremes were exceptionally fast and helped Athens achieve dominance in the field of naval warfare. This detail from the rim of a kylix (drinking cup) depicts a galley with oarsmen. Sixth century BCE. Erich Lessing/Art Resource, NY

Starting in the fourth century BCE, ships were equipped with catapults, large wooden machines with a swinging arm that could hurl stones and sink an enemy ship before it could ram. This development triggered an arms race among the Greek city-states. They built ever-larger warships to carry more powerful catapults: first *quadriremes*, then *quinqueremes* with up to 286 rowers, and finally a behemoth pulled by 4,000 rowers with up to 10 on each oar. Such enormous craft were hard to maneuver and soon became obsolete. By the second century BCE, the Romans, whose strength was their infantry, returned to large but simple ships equipped with a hinged and spiked gangway that could be lowered onto the deck of an enemy warship, allowing marines to board and capture it in hand-to-hand combat.

The first craft in the Indian Ocean and the western Pacific was the dugout canoe. Some time in the second millennium BCE, people originating from Taiwan and the Malay Peninsula began attaching an outrigger, a very thin hull attached to the side of the main hull of their canoe, thereby making it less likely to capsize. By 2000 BCE, their navigators had settled Melanesia and New Caledonia. Over the next few hundred years, their descendants mastered the art of navigating the open sea for days at a time, venturing ever eastward into the Pacific and settling the islands of Polynesia, a triangle in the Pacific between Hawaii in the north, Easter Island in the southeast, and New Zealand in the southwest. Captain James Cook, the first European to explore Polynesia in the eighteenth century, was surprised that the inhabitants of those islands thousands of miles apart spoke closely related languages. "It is extraordinary," he wrote, "that the same Nation could have spread themselves over all the isles of this Vast Ocean from New Zealand to this [Easter] Island which is almost a fourth part of the circumference of the Globe."[7] When he visited Polynesia, the people he met were Stone Age farmers and fishermen who went out in frail-looking open outrigger canoes and seldom ventured far from land. There seemed to be no contact between distant archipelagoes. How, then, did their ancestors settle the thousands of islands that comprise Polynesia?

In 1976, a group of Polynesians sailed a reconstructed outrigger canoe called *Hokule'a* from Hawaii to Tahiti using traditional navigation techniques. What their experiment showed is that the ancestors of modern Polynesians were not blown off course by storms, nor did they drift helplessly before the prevailing winds until they ran into an island by pure luck, as their detractors had claimed. Instead, they navigated by noting the position of the sun during the day and the moon and stars at night. On cloudy days, they navigated by feeling the swells of the ocean and by noting the color of clouds and the flights of birds. The intrepid sailors of the *Hokule'a* had revived the ancient art of navigation used by Polynesians for more than 3,000 years and almost forgotten in the age of steamships and precision instruments.

At the same time, Malay sailing canoes were crossing the Indian Ocean. By 1 CE, Malays had settled the large island of Madagascar. Here, as in the western Pacific, they learned to use the monsoons, winds that blow toward Asia in the summer and away from Asia in the winter. Like the sailors of the Mediterranean, they could not sail against the wind, but they knew that the wind would turn, bringing them home if they did not find land.

Although the great empires of Eurasia are famed for their architecture and civil engineering, they also made contributions to the mechanical arts by dividing work into specialized tasks. The Greek historian Xenophon, who traveled in the Persian Empire around 400 BCE, noticed this phenomenon:

> In small towns the same workman makes chairs and doors and plows and tables, and often the same artisan builds houses. . . . And it is of course impossible for a man of many trades to be proficient in all of them. In large cities, on the other hand, inasmuch as many people have demands to make upon each branch of industry, one trade alone, and very often even less than a whole trade, is enough to support a man: one man, for instance, makes shoes for men, and another for women; and there are places even where one man earns a living by only stitching shoes, another by cutting them out, another by sewing the uppers together, while there is another who performs none of these operations but only assembles the parts.[8]

Such craftsmen may have used simple tools, but their specialization did result in more efficient and productive work.

The tasks that women performed did not change much in the period of the classical empires, but the division of labor between the social classes increased markedly. The Roman farmer Columella, who lived in the first century CE, described the changes in these terms:

> Nowadays when wives so generally give way to luxury and idleness that they do not deign to carry the burden of manufacturing wool, but disdain clothing made at home, and with perverse desire are pleased with clothing that costs great sums . . . we have to appoint stewardesses on the farm to perform the offices of matron.
>
> On rainy and cold days when the slave-women cannot do farmwork in the open, let the wool be ready-combed and prepared beforehand so that they may be busy themselves at spinning and weaving, and the

steward's wife may exact the usual amount of work. It will do no harm if clothing is spun at home for the stewards, overseers and the better class of slaves so that the owner's accounts may bear the less burden.[9]

During the Han dynasty, Chinese crafts were the most highly developed in the world. Their iron makers achieved temperatures high enough to melt iron, thereby allowing them to cast the molten iron into molds and mass-produce items such as pots and sickle blades that were difficult or impossible to make by hammering wrought iron. The Chinese were centuries in advance of other civilizations in other fields as well. They began glazing pottery—that is, applying salt or lead glazes to the surface of a pot after it was fired and then firing it again to give it a smooth waterproof surface. For centuries thereafter, Chinese ceramics—called *china* in English—were so clearly superior to all others that they were exported all over the world.

The first literate Chinese wrote on slivers of bone or bamboo sewed together, using a sharp instrument. It was during the Han dynasty that craftsmen found a way to make paper out of silk and later out of rice straw or wood pulp. Paper was first used to wrap fragile items and later for writing. Paper making encouraged an outpouring of literary works as well as official documents. Paper was even used to print money.

The Chinese also invented the horse collar. Unlike the throat-and-girth harness used in western Eurasia, the collar allowed horses to pull heavy loads without choking. This device benefited wealthy Chinese who rode in horse-drawn carts, but it was little used in agriculture, for arable land was too valuable to use for pastures. In their wars against the nomadic tribes of Mongolia, the Han armies devised the crossbow, a powerful bow that used a crank to pull the bowstring and a trigger to release the arrow.

The Greeks and Romans, contemporaries of the Han dynasty, also achieved some notable successes in two of the mechanical arts: heavy weapons and water power. Their catapults were large crossbows on pedestals that used twisted bundles of sinew or hair to store energy and then released them to hurl stones up to 80 pounds or huge arrows with iron points. They were used to batter down the walls of enemy cities. Mediterranean states also used battering rams, some mounted on wheels and covered with a roof to allow the soldiers to approach the gates or walls of a besieged city. In later centuries, every Roman legion was equipped with catapults, battering rams, and other heavy equipment.

Mechanical ingenuity was also evident in the development of water-wheels. The Vitruvian waterwheel, named after the engineer Vitruvius, was a vertical wheel turned by flowing water and geared to turn millstones and grind grain. It was inspired by the *saqiya*, a Persian wheel that used the energy of a stream to lift pots of water for irrigation. A seven-foot undershot wheel—that is, one in which the water pushed the lower portion of the wheel—could grind as much grain in an hour as 80 humans. This reduced the need for women to grind grain at home, permitting them to do other work such as spinning or weaving.

During the first centuries CE, the Romans built waterwheels wherever there were streams or aqueducts furnishing water power. The one at Barbegal in southern France, built in 310 CE, used 16 overshot wheels in a row. Water flowed over one wheel, turning it, then over the next wheel, and so on. It was capable of grinding three tons of grain an hour, enough to feed 80,000 people. In other words, it was an establishment of the kind that would not be seen again in that part of the world until the Industrial Revolution of the eighteenth century.

The age of the great Eurasian empires marked a radical departure from the more conservative river valley civilizations that preceded them. Before about 1500 BCE, there had been a rough balance between the river valley civilizations and the Neolithic farmers and herdsmen who inhabited the surrounding regions. After ca. 1500 BCE, however, three technological innovations—iron, domesticated horses, and wheeled vehicles—upset this equilibrium, leading to an era of prolonged warfare that spread over much of Eurasia and North Africa. The empires that emerged from these wars ruled far more people over much larger areas than had ever been possible before. They devoted considerable resources to building great cities, canals, and other monumental public works projects and to developing new weapons and warships.

In spite of their wealth and power, the new empires of Eurasia brought few benefits to the majority of their inhabitants, for their wealth came from the work done and the surpluses created by women, poor peasants, and slaves. Nor were they more secure than their predecessors, the river valley civilizations; in fact, they were less so. For the same technologies on which their security rested—iron, horses, and chariots—were also available to people who lived outside the reach of imperial control. To the peoples of Rome, China, Persia, and India, outsiders who attacked and invaded their empires were known as "barbarians." New technologies brought new dangers as well as benefits to the people who possessed them.

CHAPTER 4

Postclassical and Medieval Revolutions (500–1400)

O n his visit to China in the late thirteenth century, the Italian traveler Marco Polo was especially impressed by "the noble and magnificent city of Kin-sai, a name that signifies 'the celestial city,' and which it merits from its preeminence to all others in the world, in point of grandeur and beauty, as well as from its abundant delights, which might lead an inhabitant to consider himself in paradise."[1] He described Kin-sai (now Hangzhou) as 100 miles in circumference, with a population of 1,600,000 families. Twelve thousand bridges spanned its many canals. The markets in its ten main squares, each half a mile long by half a mile wide, carried more products than Polo had ever seen. Its streets were paved with stones and bricks, as were all the principal roads throughout the province. To his amazement, he recalled in his travel memoir, "The inhabitants of the city are idolaters, and they use paper money as currency. . . . The greater part of them are always clothed in silk."[2] Although Marco Polo often exaggerated, there is no question that Hangzhou was the largest city in China, many times larger, wealthier, and more magnificent than his hometown of Venice, at that time the richest city in Europe. Marco Polo's amazement was that of a visitor from a backward place to the most advanced and developed region in the world.

Since the time of the Sumerians, the wealth of any country was based on its agriculture. It is the surplus produced by farmers, over and above their subsistence needs, that supported governments, trade, and urban life. The nature of a society and the size and prosperity of its cities were the result of the productivity of its agriculture. The centuries after 500 CE saw remarkable advances in agriculture in three areas of the world: China, the Middle East, and Europe.

For more than 2,000 years, most of the Chinese people had lived in the north, near the Yellow River, where they grew wheat and millet on rain-watered land. Over time, they developed a number of farm implements that made their agriculture more productive. The mold-board plow consisted of a vertical coulter or blade that sliced the sod, a horizontal plowshare that cut the soil at the grassroots, and a curved moldboard that turned it over. The Chinese also devised harrows that broke up the clods of earth and seed drills that spaced the seeds and covered them with manure. Yet in the cold climate of northern China, farmers could grow only one crop a year.

During the early first millennium CE, Chinese people slowly mi-grated south into the Yangzi River Valley and into the hills of south-ern China, where they could grow rice, which requires warmth and water. The tenth century was a time of troubles, during which the north was periodically invaded and harassed by nomadic warriors from the grasslands of Mongolia. Unable to prevent these barbarian incursions, the rulers of the Song dynasty encouraged the Chinese people to move south. To improve their agriculture, the Zhenzong emperor imported 30,000 bushels of rice seeds from Vietnam in 1012. This variety, called Champa rice, ripened so fast that it was possible to harvest two crops a year, even three in tropical latitudes.

Champa and other varieties of rice grew only in water. The land, or rather the mud in the rice paddies, had to be plowed several times with a moldboard plow pulled by a water buffalo or several men. Seedlings were prepared in special boxes and then transplanted by hand into the paddies. The water level had to be carefully monitored and regulated, and the fields had to be weeded frequently with a weeding rake. For the first few years, rice paddies had to be fertilized with ashes, river silt, and urban refuse. After that, the algae that grew in the water absorbed nitrogen from the air, making the paddies more fertile with each passing year.

Once flat areas along riverbanks had been occupied, new migrants sought other land to till. Marshy areas such as the delta of the Yangzi had to be drained and protected by dikes. Hillsides were reclaimed by terracing narrow but perfectly level fields that followed the contours of the land and were held in place by stone walls. Wet-rice fields required intricate water-control systems, not only ditches and canals but also devices to lift water into or out of the paddies. The well-sweep, a coun-terbalanced pole with a bucket on one end and a weight on the other, had been in use for centuries; the *noria*, a waterwheel powered by the current of a river, was a Middle Eastern invention; but the "dragon-backbone machine," a chain of buckets turned by a man on a tread-mill, was a recent Chinese invention. The government was responsible for flood control on the major rivers, but most irrigation and drainage works were local projects, built with shovels and baskets under the supervision of local officials.

The Song and later governments contributed to the opening of new lands to rice agriculture with a number of measures besides providing seeds. They made land grants to wealthy men who supplied tools and seeds to their tenant farmers and provided low-interest loans and tax rebates on newly reclaimed land. To persuade landlords to introduce new techniques to their tenants, they appointed "master farmers" to act as agricultural extension agents and published practical manuals on farming and water control, such as the one in which the fourteenth-century agronomist Wang Zhen described a new weeding device

> shaped like a wooden patten, a foot or so long and roughly three inches broad, spiked underneath with rows of a dozen or so nails . . . the farmer stirs together the mud and weeds between the rows of crops so that the weeds are buried in the mud . . . In certain areas I have seen farmers weeding their fields by hand, crawling between the crops on their hands and knees with the sun roasting their backs and the mud soaking their limbs—a truly pitiable fate, and so I have described the hand-harrow here in the hope that philanthropists will disseminate their use.[3]

The productivity of the land depended mainly on the amount of labor that went into it. If carried out with care and diligence, wet-rice cultivation provided the highest yields—that is, the most food per acre—of any form of agriculture, one that could support the highest population densities in the world. In one part of Sichuan in central China, 5 million people farmed an area of 40 by 50 miles. The productivity per person, however, was very low, which meant that farmers worked very hard and earned very little. Among them, women were even worse off than men, as Wang Mian, a fourteenth-century writer, explained:

> During the daytime they labor in the fields, following their husbands, and then, during the night, they spin hemp. They do not go to bed. The hemp they have spun is made into cloth which is used to meet that part of the official tax levied in cloth. The rice won by their toil in the fields goes into official granaries. So long as payment to the officials have not been met completely, they suffer from pent-up depression.[4]

Not all farming was devoted to rice, of course. Growing mulberry trees to feed silkworms was the major activity of some regions, and others produced tea, bamboo, cotton, oil seeds, citrus, and other fruits. Along with agriculture came many crafts such as weaving silk, preparing tea leaves, or making charcoal, paper, lacquer wares, or rice wine.

Throughout the Song domains, there was an upsurge in craft production and a growth in the market economy. The population of China doubled. By 1080, twice as many Chinese lived in the south as in the north of the country, many of them in large cities such as Hangzhou.

Meanwhile, at the western end of Eurasia, the western Roman Empire collapsed under the onslaught of the Germanic barbarians. Cities were abandoned, the population declined, and many of the Roman achievements in architecture and civil engineering were lost. The eastern half of the empire prospered for another thousand years, however, first under the Byzantine Empire and then under the Arabs. The most momentous change occurred when the people of Arabia, stirred to action by the Islamic religion, conquered North Africa and Southwest Asia between the seventh and the ninth century. The Arabs were eager to learn and to exchange ideas with the peoples they encountered. They traded actively and built great cities, such as Baghdad in Iraq, Cairo in Egypt, and Cordoba in Spain.

The Arabs, a desert people, were especially interested in irrigation and water control. In regions that had long practiced water control, such as Mesopotamia and Egypt, they rebuilt, improved, and extended existing irrigation and drainage systems. They borrowed the idea of the *qanat*—a tunnel to carry irrigation water over long distances—and spread it from Persia to other parts of the Middle East. They also improved the *saqiya*, a chain of pots to lift water, using the power of a blindfolded animal turning in a circle.

Arab traders brought new crops from China, India, or Southeast Asia and spread them throughout their empire. From China, they brought oranges and lemons; from India, sugar cane and cotton; and from other parts of Asia, they brought melons, asparagus, bananas, spinach, and other delicacies. During the height of their empire, from the mid-seventh to the mid-eleventh century, the Arabs surpassed the level of prosperity and technological sophistication that the Middle East and North Africa had known in Greco-Roman times. Only China was more developed.

The prosperity of the Middle East under the Arabs rested on several other technological advances, some of them inherited from China, such as paper and the compass, and others developed by the Arabs themselves. They paid special attention to astronomy, a science that was vital to navigation. They also excelled in mathematics. Arab mathematicians transmitted our "Arabic" numerals and the concept of zero from India to the Middle East and from there to Europe. Our word *algorithm* is named after the Arab mathematician Al-Khwarizmi, whose book *Kitab al-Jabr* gave us the word *algebra*. The origins of chemistry can be traced

The Arabs, coming from a dry land, eagerly built or rebuilt irrigation systems throughout their empire. This fourteenth-century waterwheel, which is still in use in Hamah, Syria, lifted the water to the aqueducts that provided water for drinking and irrigation. DeA Picture Library/Art Resource, NY

to experiments in *alchemy*, another word derived from Arabic. Their knowledge of medicine advanced far beyond that of the Greeks. The physician Abu al-Qasim al-Zahrawi wrote a medical encyclopedia that was used in medical schools for several centuries. Others pioneered the use of anesthesia in surgery and developed many drugs based on plants. During the Middle Ages, Christian European scholars came to Spain— then under Muslim rule—to translate the works of Arab scientists and thereby transmitted to Christian Europe many scientific and technical discoveries of the Greeks, Indians, and Chinese.

In European history, the half-millennium from 500 to 1000 is often called the Dark Ages. As barbarian tribes invaded the western Roman Empire, the imperial government collapsed, and trade and manufactures declined. Cities fell into ruin; compared to Baghdad with its million inhabitants or Cordoba with almost half a million, the largest cities in western Europe were mere towns: Rome, once a metropolis of a million, shrank to 35,000 inhabitants; Paris had 20,000 and London, 15,000. Literacy almost disappeared as well. Charlemagne, king of the Franks from 768 to 814, could not read or write. European society became decentralized, almost down to the village level.

Medieval surgical instruments such as scalpels, lancets, and forceps were made of bronze, iron, or silver in shapes that a modern surgeon would recognize. This image is from an Arabic manuscript attributed to Abu al-Qasim Khalaf ibn al-Abbas Al-Zahrawi, known to Europeans as Abulcasis, medieval Islam's greatest surgeon. Arab physicians of that era produced numerous medical treatises and encyclopedias for the instruction and guidance of the profession. Bildarchiv Preussischer Kulturbesitz/Art Resource, NY

Most Europeans were farmers producing for local consumption. Above them were knights, warriors who owed allegiance to higher lords and kings but whose main occupations were protecting their peasants and fighting one another. The clergy formed a third group; some could read and write in Latin, but others were illiterate. Yet this society produced a technological revolution in agriculture that set the stage for the modern world. This revolution was based on three innovations: the moldboard plow, three-field rotation, and much more effective use of the horse.

The moldboard plow was a much heavier and more complicated device than the scratch plow used by the Romans. Such a plow, known in China since the third century, was independently invented in Europe and first used in the sixth century by Slavic peasants in eastern Europe, from where it gradually spread westward to northwestern Europe. Because it was heavy and dug deeply into the soil, the new plow required very different farming methods from those of the Romans. It took a

The moldboard plow, invented in the early Middle Ages, turned over the soil, letting excess water drain off and killing weeds. It allowed agriculture to spread to areas of northern and western Europe that the Romans, who only used a simple scratch plow, could not colonize. Scala/Art Resource, NY

team of eight oxen to break new land, and even well-tilled land required four oxen, more than the average farmer could afford. Hence, farmers had to join together to plow their fields. The plow broke the soil so well that farmers did not need to plow a second time at a right angle, as they did with a scratch plow that only cut a thin furrow in the topsoil. Because a team of oxen was so hard to turn, farmers preferred fields that were long and narrow rather than square. The great advantage of the moldboard plow was that it could be used in the heavy wet soils of northwestern Europe, especially the fertile river bottoms that the Romans had never mastered.

Three-field rotation contributed as much to European agriculture as the heavy plow. Traditionally, farmers had planted only half their land in a given year, leaving the other half fallow to allow it to recover its fertility. The following year, they switched fields. During the early Middle Ages, European farmers learned that they could grow wheat, rye, or oats on one-third of their land and peas or other legumes on another third while leaving the last third fallow. The nitrogen that the legumes took from the air, along with animal manure, fertilized the soil. Compared to the old two-field rotation, this method increased the productivity of the soil by one-third to one-half with no additional labor. It also reduced the risk of crop failure, provided more protein in the diets of humans, and produced oats for horses.

Before the Middle Ages, horses could not be used for farming or pulling heavy loads because the throat-and-girth harness choked them and caused them to rear up if the load was too heavy. In the third century BCE, the Chinese invented the horse collar that allowed the horse to push with its shoulders and did not interfere with its breathing; an animal could now pull four or five times more than with a throat strap. The same device was used in Tunisia in late Roman times to harness camels to carts and plows; that is probably where Europeans got the idea. By the late ninth century, horses with collars were pulling plows in Scandinavia, and by 1100, they were common throughout western and central Europe.

Another invention, the nailed horseshoe, permitted horses to be used for heavy pulling in areas of moist climate like western Europe, where horses' hooves would otherwise get soft and quickly wear out. This invention appeared in Siberia in the ninth or tenth century and spread to China and Europe by the eleventh century. By then, every village had a blacksmith, and iron became much more common than it had been in Roman times.

Finally, Europeans bred larger and stronger horses than those found in central and eastern Asia, horses well suited for heavy agricultural work. These horses walked faster than oxen and could work one or two hours a day longer; in short, they could do twice the work of oxen. However, they could not subsist like cattle on hay and the stubble left in the fields after the harvest but needed oats. With the three-field rotation system, it was possible to grow oats for the horses one year of three. What resulted was a complicated agricultural system integrating animals and plants, with many activities spread throughout the year. It transformed rural life, as farmers no longer had to live in isolated hamlets near their fields but could live in larger villages with a church, a blacksmith's shop, and a tavern and still get to their fields in the same amount of time. The surpluses that the new agriculture produced led to a revival of trade. As the population increased, villages turned into towns, and towns became cities; by 1300, Paris had 228,000 inhabitants.

From 500 to 1400, human ingenuity did not limit itself to improving agriculture and other civilian endeavors but also transformed, once again, the art of war. As had often happened in the past, the advantage on the battlefield shifted back and forth between the infantry and the cavalry, and in the towns, it shifted between defensive walls and offensive weapons. Among the innovations that marked this period were the armored knight and siege machinery.

Horses had been known throughout Eurasia and North Africa for thousands of years. In the grasslands of Russia and central Asia, nomads used them for cattle herding, raiding, and warfare. In settled agricultural regions, horses were costly luxuries, long used to carry messengers and for ceremonial occasions. It took several technical advances to make them useful in other areas.

In the first century BCE, Persian landowners created a new breed of large horses strong enough to carry a fully armed knight. In 101 BCE, the Han Emperor Wudi sent an expedition to central Asia to bring back what the Chinese called "blood-sweating horses." Such powerful beasts could not survive on grass alone but needed oats and alfalfa, crops that competed with human needs.

With the end of the Han dynasty in 220 CE and the fall of the western Roman Empire in 476, warfare became endemic throughout Eurasia. New weapons and ways of fighting gave decisive advantages to innovators but never for long. Unlike agricultural techniques, which are specific to particular environments, military technology diffuses rapidly from one people to another, as warriors imitate their enemies. The classic empires had all relied on foot soldiers who could hold off cavalry attacks by using spears or bows and arrows because riders sat precariously on their mounts and needed one hand to hold on, and cavalrymen often dismounted when they reached the battlefield.

What now gave cavalrymen an advantage over foot soldiers was a simple device, the stirrup. Sitting on a saddle with his feet firmly planted in his stirrups, a rider no longer risked falling off. In effect, he became one with his horse. Toe stirrups were known in India in the second century BCE, and foot stirrups appeared in northern Afghanistan in the first century CE. From there, they spread to China, where they were common by the fifth century. However, the stirrup gave the advantage to the long-standing enemies of the Chinese, the nomadic Mongols, who spent much of their lives on horseback. Being expert archers, they could now shoot arrows in midgallop, overwhelming Chinese infantrymen and cavalrymen alike.

By the early eighth century, the stirrup had reached Europe, where it caused a revolution in warfare. Charles Martel, leader of the Franks, was the first European leader to realize the potential of equipping cavalrymen with a heavy lance held under the arm. With the lance firmly attached to the rider and the rider attached to the horse by stirrups, the lance transmitted not only its own weight but also that of the rider and the horse. It was therefore a much more powerful weapon than the javelin or spear that was thrown. For several centuries, only another knight or the walls of a city could stop a lance-bearing knight. To protect themselves from each others' lances, knights had to wear heavy suits of armor and carry heavy shields on their left arm. Later, even the horse was armored. Horses were specially bred to be large and strong to carry the weight of the rider and his armor. Every knight had to own a suit of armor and at least two battle horses (one to ride and a spare). Knights employed squires to care for the horses and tenant farmers to grow the oats the horses needed to stay strong. They practiced shock combat from childhood, engaging in mock battles called tournaments with other knights. Warfare became the main activity of professional warriors, and kings granted lands and villages to the best warriors in exchange for loyal service and prowess on the battlefield in time of war. It was with just such an army that William of Normandy crossed the English Channel in 1066. At the battle of Hastings, King Harold's English warriors dismounted when they reached the battlefield, only to be impaled by the lances of William's knights.

In response to the appearance of heavily armored knights, armorers developed the crossbow. The longbow traditionally used by European archers was a long piece of an elastic wood such as yew with a bowstring that was pulled back by the archer and, when released, imparted its energy to the arrow. In contrast to this simple device, the crossbow was so hard to bend that the string could not be pulled by the archer's arm but only by turning a crank. Although a crossbow required less strength or skill on the part of the archer than a longbow did, it was difficult to make, for it required several iron parts. In battle, it hurled a heavy arrow that could penetrate a suit of armor and was so powerful that in 1139 Pope Innocent II is said to have banned its use against Christians. Despite this prohibition, by the fourteenth century, the crossbow had canceled the advantage that the armored knight had long enjoyed on the battlefields of Europe.

Not all fighting took place on the battlefield. Towns and cities built walls to protect themselves against marauding warriors. To attack fortresses and fortified towns, the Greeks and Romans, like the Chinese and Indians, had long used torsion catapults, large machines that stored energy by twisting fibers or hair and released it suddenly to throw a heavy stone. However, such devices were of limited use in wet climates like those of southern China and western Europe, where hair and fibers lost their elasticity. There, engineers developed catapults that operated on new principles. The first was the swing-arm catapult, in which a long beam pivoted on a frame, with a projectile at one end and ropes pulled by men at the other. By 1200, they were common in Europe, quickly displacing torsion catapults. Even more powerful was the trebuchet, or counterweight catapult. This was a large swing-arm catapult with a hinged counterweight on one end that could launch a boulder weighing a ton or more, compared to the 40- to 60-pound stones torsion catapults could throw. It was probably of Middle Eastern origin, for it first appeared in China in the 1270s under the name Muslim stonethrower and was used by the Mongols in their conquest of that country. It was also widely used in Europe and the Middle East.

The trebuchet was an example of the advances in mechanical engineering that took place after about 500. Throughout Eurasia, people became more familiar with mechanisms that had moving parts, such as waterwheels, mills, spinning wheels, and clocks. Although they were often made of wood, many of these devices were more complex than those of any previous civilization.

Since the Han dynasty, the Chinese had used waterwheels to power the bellows that provided a blast for forging iron and for hulling rice, as well as millstones to grind grain. In addition to the more common waterwheels that turned in a vertical plane, the Chinese also developed horizontal wheels in which the water flowed through the wheel—as in the turbines used in hydroelectric dams today—instead of over or under it.

The Islamic world also knew waterwheels, but more sparingly, for there were fewer streams in the dry regions of North Africa and western Asia. Yet there seems to have been a tidal mill in Basra, on the Persian Gulf, around the year 1000 and floating waterwheels in Baghdad on the Tigris River. During the first three centuries CE, the Romans had built mills powered by waterwheels, but these were unusual at the time. After the eighth century, water-powered mills proliferated throughout western Europe, where a rainy climate created an abundance of streams. At first, they were used only for grinding grain, but after ca. 1000, waterwheels were powering sawmills, fulling cloth, hammering iron in forges, crushing ore, and many other processes. The *Domesday Book*, a census of England ordered by William the Conqueror in 1086, counted 5,624 mills in 3,000 communities, one for every 50 households. These complex machines moved by inanimate energy were becoming part of the daily experience of almost all Europeans.

Streams and rivers often froze in the winter or dried up in the summer. In certain locations, the wind was a more reliable source of energy. The first windmills, built in Persia and Afghanistan in the tenth century, had a vertical axle and curved vanes. Vertical mills spread to other parts of the Middle East and to China but were never very common. The European windmill, in contrast, had a horizontal axle at the top of a tower and gearing to transmit energy to ground level. The first such windmill was built in 1185 at Weedley in Yorkshire, England. The idea spread quickly; within seven years, there were windmills in much of Europe and as far east as Syria. So important were they that Pope Celestine III imposed a tithe on windmills. Soon windmills were grinding grain, pumping water, and performing other duties in windy regions such as Holland. Because of the high cost of labor in Europe, they were more popular there than in China or the Middle East, where such tasks were performed by servants and women.

In textile machinery, China and India were several centuries ahead of the rest of the world. India specialized in fine cotton cloth, which it exported to other countries. The spinning wheel allowed the spinster to

Winding devices were originally used in China to convert long silk fibers into threads, and the technology was later adapted to cotton. Spinning wheels were widely used in China by the eleventh century. Art and Architecture Collection, Miriam and Ira D. Wallach Division of Art, Prints, and Photographs, The New York Public Library, Astor, Lenox and Tilden Foundations

work sitting down; its more even speed of rotation produced yarn faster and more evenly than the ancient spindle. It originated in India in the eighth century CE; by the twelfth century, it had spread to the Middle East, Europe, and China.

Spinning wheels were handicraft devices used by women in their homes. In China, spinning and weaving were widespread because taxes were paid in kind: each peasant household owed the government two bolts of plain cloth and two bushels of grain a year. For more elaborate weaves such as brocades and satin, the Chinese developed complex industrial machines for use in large urban workshops. The silk-reeling machine, perfected in the eleventh century, drew several filaments simultaneously from cocoons immersed in boiling water. In his *Treatise on Agriculture* published in 1313, the agronomist Wang Zhen described a spinning machine for hemp with 32 spindles powered by a waterwheel; it was, he said, "several times cheaper than the women workers it replaces."[5] Labor-saving devices such as this increased the productivity of labor but at the expense of workers in the crafts thus rendered obsolete.

The wheelbarrow was another simple device that saved a great deal of labor on farms and construction sites. As with so many other inventions, the wheelbarrow originated in China, where it was known since the second century. The concept spread to Europe in the twelfth century, perhaps via the Middle East.

One area in which Europeans surpassed other civilizations was in the measurement of time. The first instrument to measure time was the Egyptian *gnomon*, a vertical stick that cast a shadow that told the time of day. The Greeks and Romans built sundials that divided the day into 12 equal segments. In the thirteenth century CE, the Arab Abu al-Hassan al-Marrakushi invented a sundial that divided the day into equal hours. Sundials only work on sunny days, of course. To tell time at night and in cloudy weather, peoples used *clepsydras*, or water clocks. The ancient Babylonians and Egyptians made vessels with a hole in the bottom; as the water dripped out, they could read off the time by the level of the water that remained. In 807, the caliph of Baghdad Harun al-Rashid gave the Emperor Charlemagne an elaborate water clock that turned gears, animated figures of horsemen, and rang bells. In 1086–1094, Su Sung had a water-powered mechanism built for the emperor of China that told time and showed the movements of the sun, the moon, and the stars. Water clocks stopped when they froze, however, and it was difficult to put them atop a tower because of the need to supply them with water.

Medieval Europeans, who lived in a region where it was often cloudy and sometimes froze, were the first to build weight-driven mechanical clocks. What made the mechanical clock possible was the invention in the late thirteenth century of the escapement, a device that transformed the accelerating movement of a falling weight into a series of short movements of equal length. Early clocks were so inaccurate that they only had hour hands. But they could be placed on a tower, to the greater glory of the local prince or town fathers, and they could be made small enough to fit inside the houses of wealthy people.

In the manufacture of iron, as in so many other areas, Song China was far in advance of the rest of the world. Since the Han dynasty, the Chinese had known how to raise the temperature of a furnace, using water-powered air pumps, to turn ore into raw (or "pig") iron. Unlike the spongy bloom made elsewhere that had to be hammered into wrought iron, molten pig iron could be cast into molds to make pots, bells, plowshares, arrow- and spearheads, even statues.

Chinese iron workers were also the first to use coal as fuel in iron furnaces, a necessity in places where wood was in short supply. Most coal, however, produces fumes that damage the iron, so iron makers could only use a very pure coal called anthracite. In the early eleventh century, they found a way to use ordinary coal by baking it first to remove the impurities, leaving coke, a relatively clean form of carbon.

In the tenth and eleventh centuries, Song China was at war with the Khitan and later the Jürchen, nomadic warriors from the steppe lands to the north. The government assembled the largest army ever seen until that time, with 1,250,000 soldiers. To equip them, it operated a huge armament industry that produced 16 million arrowheads and 32,000 suits of armor a year. In 1078 alone, the blast furnaces of Hebei-Henan produced 125,000 tons of iron, a figure not surpassed anywhere in the world until the nineteenth century and probably more than the rest of the world put together. Some of the ironworks were huge industrial establishments employing hundreds of workers. Iron was so cheap that the Chinese even built suspension bridges with iron chains supporting the roadway.

Despite their large army, the Song were defeated in 1126 and lost control of northern China, including the ironworks of Hebei-Henan. For a century and a half, China was divided in two. The Song ruled the south, a land of mountains, rivers, lakes, and marshes, where they protected themselves with a fleet of riverboats carrying archers, soldiers with long spears or pikes, and catapults. Meanwhile, the Jürchen ruled the north, taking the name Jin dynasty. When the Song were attacked by Jin warriors from the north, a general named Li Kang devised wheeled vehicles armored with iron plates for defense against the enemies' arrows.

Much of the iron produced under the Jürchen was sold to the Mongols, nomadic herdsmen and warriors from the grasslands of Mongolia. Like the Jürchen, the Mongols employed Chinese experts to set up smelters and produce iron weapons. Armed with iron spears and arrowheads, the Mongols defeated the Jürchen in 1226 and the Song in 1277, and they became the rulers of China and much of the rest of Asia as well. In the turmoil of war, iron production dropped to a quarter or less of its peak under the Song and did not recover for eight centuries.

While the Chinese excelled in mass-producing inexpensive iron, craftsmen in the Middle East surpassed the Chinese in the quality of their swords and daggers. Their damascene blades, named after the city of Damascus in Syria, were made of a special kind of steel called *wootz* imported from India. Ingots of wootz the size of hockey pucks were put in clay crucibles along with charcoal and leaves and then heated to around 1,200 degrees Celsius. This gave the metal a carbon content of 1.5 percent, making it hard but brittle. To make the metal resilient, the puck was then removed and forged—that is, hammered and allowed to cool slowly and then reheated and hammered again up to 50 times, until it was shaped into a blade that was tough and flexible, yet so sharp that it could slice a silk handkerchief in midair. Indian wootz contained trace elements (less than 0.03 percent) of vanadium or molybdenum, two rare metals that caused the steel to congeal in alternating bands of darker and lighter color rather than mixing uniformly. This imparted a distinctive pattern for which damascene blades have been admired for centuries but never imitated.

European iron workers at the time produced far less iron than the Chinese and of much lower quality than Middle Eastern smiths. Nonetheless, iron was more common during the European Middle Ages than it had ever been in Roman times. Like so much else in Europe, iron making was decentralized, with iron furnaces in every region and blacksmiths in every village, forging horseshoes, plowshares, sickles, spades, other tools, and the occasional suit of armor.

With the increased economic activity and military campaigns came an increase in travel and trade over long distances, hence the development of better means of transportation. Although a great deal of travel and trade went by sea (as we will see in chapter 5), inland transportation also changed substantially. The four civilizations described in this chapter—Song China, the Arab Empire, western Europe, and the Incas—devised very different methods of transporting goods inland.

China has several rivers that flow from west to east but none that flow north or south. From the time of the first emperor, Shi Huangdi (260–210 BCE), the Chinese attempted to link their river systems with canals. During the Sui dynasty (581–618), they built the Grand Canal linking the Yellow River in the north with the Yangzi River in central China. Boat traffic on the Yangzi increased along with the population of the region. When they ruled all of China, the Song built canals from their capital, Kaifeng, north to the iron region of Hebei-Henan and south to the Yangzi; during the Southern Song period, they added to the canal network of the south. Their successors, the Mongol or Yuan dynasty (1260–1368), completed the 1,100-mile-long Grand Canal in 1327, allowing the transport of rice from Hangzhou and the Yangzi Basin to the Mongol capital at Beijing.

Not all rivers and canals were at the same level, and until the twelfth century, boats had to be hauled up or down ramps from one canal to another. The invention of locks with double gates allowed water from the higher canal or river to fill the lock and lift boats in it. For those going downstream, the upper gates were closed and the lower gates opened, letting water flow out and lowering the boats in the lock.

In the Middle East, much trade went by ships on the surrounding seas. Except on the Nile, the Tigris, and the Euphrates, where boats had not changed much since ancient times, the lack of navigable rivers and canals limited overland transportation to high-value items such as spices, incense, and fine textiles and handicrafts. The Romans, like their predecessors the Greeks and Persians, had used carts pulled by oxen, horses, or donkeys as well as horse-drawn war chariots. For both military and commercial reasons, they built a network of roads, many of them paved with stones. But such roads were very expensive to maintain. With the domestication of camels, which could carry heavier loads and survive in dry and broken terrain, the need for roads declined.

The camel was first domesticated in southern Arabia in the third millennium BCE. By the second millennium BCE, it was known—but rare—in Egypt and Mesopotamia. Gradually, its use spread as camel-breeding desert dwellers from Arabia grew in numbers and power. Desert traders used the South Arabian saddle, placing baggage over the animal's hump and shoulders, while the rider sat behind the hump and controlled the camel's movements with a long stick. Although good for transportation, this system made the camel useless for combat, for the rider could easily be knocked off the camel.

The North Arabian saddle, invented sometime after 500 BCE, rested on the shoulders and haunches of the animal, straddling its hump, and allowed it to carry a load of up to 300 pounds. From this position, a rider could wield a sword or spear and control the animal's movement with his feet. Thus mounted, the desert tribes of Arabia became very powerful. They could control and tax the merchants' caravans that crossed the desert and attack towns and farmlands beyond the desert. Camels contributed to the Arabs' rise to military and economic power. As their power grew, so did the use of camels. After the Arabs had conquered the Middle East in the seventh and eighth centuries CE, the camel, able to survive for weeks without eating or drinking, became the main means of transportation in southwestern Asia. Starting in 753, it allowed merchants to cross the Sahara Desert, bringing the Sudanic region of Africa into contact with North Africa and the Middle East. In places outside the Arab Empire where camels were common, such as Anatolia, northern India, central Asia, and western China, wheeled vehicles and camels coexisted, and camels were sometimes harnessed to carts. In colder regions, the two-humped or Bactrian camel with short legs and thick fur carried most of the trade along the Silk Road between the Middle East and China via the oases of central Asia.

In Europe, the revolution in transportation came, in large part, from the same inventions that revolutionized farming—namely, the padded horse collar and the horseshoe. Starting in the ninth century, horses were used to pull heavy loads at twice the speed of oxen. Carpenters built four-wheeled carts with swiveling front axles, brakes, and singletrees (whippletrees), a beam that equalized the load pulled by two horses when turning a corner. Such heavy wagons were a common sight by the twelfth century.

By law, the Romans had limited the weight of horse-drawn wagons to 717 pounds to prevent the horses from choking. The medieval inventions allowed horses to pull four or five times as hard, tripling the amount of cargo a team could pull. The result was a major drop in the cost of transport. In Roman times, transporting bulky goods 100 miles overland doubled their cost, but by the thirteenth century, it raised their cost by a third. By the twelfth century, Europeans were traveling in great numbers on pilgrimages to holy sites such as Rome or Canterbury in England or Santiago de Compostela in Spain. Just like the canals of China or the camels of the Middle East, horse-drawn carts contributed to regional specialization and long-distance trade throughout Europe, as it now became practical for merchants to bring valuable goods like woolen cloth and metalwork to great fairs held annually in Burgundy, Flanders, England, and Germany.

In contrast with the empires of the Eastern Hemisphere, the peoples of the Americas had no horses, wheeled vehicles, or iron of any kind before the Spanish arrived in 1492. Yet one of the states of South America, the Inca Empire, was as well organized and as devoted to massive construction projects as the Persian, Roman, or Chinese Empires had been centuries earlier. Technology is a matter of organization as well as of artifacts.

The Incas originated in the highlands of Bolivia and Peru. In the thirteenth century CE, they began conquering their neighbors. By the early fifteenth, they ruled an empire that stretched 2,700 miles from Chile in the south to Ecuador in the north. In their capital city Cuzco, in Peru, they built temples and palaces out of massive stones weighing up to 100 tons fitted together without mortar, yet so precisely carved that it is impossible to insert a knife blade between them. Among their most extraordinary constructions was a royal refuge and ceremonial center

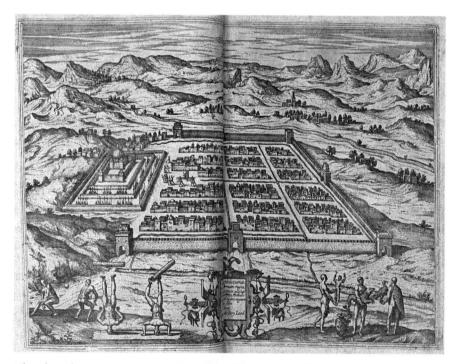

Theodore de Bry's engraving of Cuzco, published in a 1596 history of the Americas, depicts the methodical design and sturdy architecture of the city. The artist added acrobats in the foreground for the viewers' amusement. The Andes Mountains in the background, however, are real. Library of Congress

called Machu Picchu built on a mountaintop high in the Andes; when the American explorer Hiram Bingham first saw it in 1911, he wandered through "a maze of beautiful granite houses . . . covered with trees and moss and the growth of centuries. . . . Walls of white granite . . . carefully cut and exquisitely fitted together." When he came to the plaza with two temples and a large stone altar, "the sight held me spellbound."[6]

At its height, the Inca Empire contained between 6 and 10 million subjects. To control such a vast empire and bring tribute in the form of food, cloth, and metal goods to the capital, the Incas built two parallel highways, one along the Pacific coast and the other in the Andes at an average altitude of 13,000 feet, along with several short east–west roads linking the two. Since they designed their roads for pedestrians and llamas—their only pack animals—they made them as straight as possible to minimize the distance traveled. They spanned deep gorges with suspension bridges made of thick fiber cables anchored in masonry. In places, they carved six-foot-wide cuts through solid rock. Near cities, they paved their roads with perfectly fitted stone blocks; elsewhere, roads were of packed earth bordered with stones. Every few miles, there were post houses for specially trained long-distance runners. A message could travel 1,500 miles in a week, a speed comparable to that of the horseback-riding Persian or Roman messengers. Thanks to their roads and runners, the Incas were able to hold together an empire as large as any in the Eastern Hemisphere, despite the lack of large animals or navigable rivers.

Among the many different societies and political units in the world at this time, China was clearly the most dynamic and technologically advanced. Many technologies—horse collars, canals and locks, cast iron, textile machinery—originated in China and later spread to other parts of the Eastern Hemisphere.

The Arab Empire was also dynamic but in a different way. The Arabs excelled in the sciences, especially mathematics, astronomy, and medicine, in which they revived the knowledge of the classical civilizations and contributed much of their own. They revived and perfected the methods of the Greeks and Romans and introduced new techniques in agriculture, irrigation, and navigation.

The revolution in wet-rice agriculture under the Song produced astonishing yields and supported the highest population densities in the world, but it depended entirely on human labor. By improving their techniques and intensifying their efforts, more and more farmers could subsist on the same amount of land. Likewise, the Arab Empire restored the labor-intensive agriculture of ancient times. In the Middle East, as in China and South America, labor was abundant but arable land was scarce. It was the introduction of new crops and the increase in trade that contributed to the growth of markets and cities.

In comparison to these two civilizations, western Europe was backward and underdeveloped. That is why Marco Polo was amazed and delighted at the wealth and high technology that he discovered in China. Yet technologies cannot be ranked on a single continuum, for there are differences that go beyond the simple categories of "advanced" or "backward." In particular, European technology differed from those of China and the Middle East in two significant ways: sources of energy and decentralization. Europe had fewer people and more land per person available for farming than China or the Middle East. The agricultural implements that European peasants used were no more complex or efficient—often less so—than their Chinese or Middle Eastern equivalents. But European farmers relied far more heavily on animal power than any other people. In the long run, reliance on nonhuman sources of energy was the foundation for improvements in living standards.

Furthermore, much of the technology developed in China and, to a lesser degree, in the Middle East was directed and controlled from above. During the Song period, the government distributed seeds, constructed canals, published technical manuals, invested in ironworks, and encouraged innovations in many other ways. In the Middle East, likewise, the rebuilding of irrigation works, harbors, and other public works was carried out by the central or provincial governments. In Europe, in contrast, almost all the technological changes we have seen originated in local initiatives, at the level of village artisans or farmers. Although the centralization of power encouraged technological innovations and hastened their diffusion, it also meant that technological change was vulnerable, for a shift in government policy could just as easily hinder as help the process. Decentralized innovation, as in Europe, was not so likely to be stopped by a change in the policies of a government.

CHAPTER 5

An Age of Global Interactions (1300–1800)

In early 1498, four ships under the Portuguese Captain Vasco da Gama slowly made their way up the east coast of Africa. In the Kenyan port of Malindi, da Gama hired an Arab pilot to guide his fleet to Calicut in India. His small ships—only 90 feet long—must have seemed puny to the people of Calicut. Not that long before, between 1405 and 1433, Chinese fleets had sailed these same waters with dozens of huge ships, some of them 400 feet long, that would have dwarfed the Portuguese. But the Chinese left in 1433 and never returned, whereas the Europeans had come to stay.

New kinds of ships and navigation techniques changed the world by opening up communications between formerly isolated or semi-isolated regions of the earth. At the same time, three other technological changes transformed the world between 1300 and 1800: some states became what historians call "gunpowder empires" by using firearms to conquer their neighbors; crops brought from one part of the world to another allowed humans to multiply in numbers; and printing revolutionized access to information and ideas. These innovations involved more than the introduction of improved technologies, for some people adopted them faster than others; they not only gave each civilization a different character but also allowed one region to forge ahead of the others. Western Europe, which had lagged behind China and the Middle East, used techniques borrowed from these two civilizations to create global empires, first at sea and then on land. The arrival of Vasco da Gama in India marked a turning point in the balance of power in the world.

Until the mid-fifteenth century, many peoples had designed ships for particular nautical environments. Sailing to distant seas where the climate, winds, and currents were very different from those in home waters, however, required advances in nautical technology that only the Chinese possessed. During the Song dynasty (960–1279), Chinese shipbuilders began launching oceangoing ships called *junks* to compete with merchant ships from Southeast Asia and the Indian Ocean. Junks were built with a flat bottom to sail in shallow waters and keels that could be lowered for stability at sea. For safety, they were divided into separate compartments by watertight bulkheads, a feature not found elsewhere for another four centuries. Three to twelve masts carried sails stiffened by bamboo, like a fan. Junks were the first ships equipped with sternpost rudders rather than steering boards, making them easier to maneuver. They were not only larger than other ships, but they were also stronger and better able to withstand storms. By the eleventh century, such junks were taking over the trade with the Indian Ocean from the smaller ships built in India or the Middle East. The Mongols who invaded China in the late thirteenth century had thousands of ships built for military purposes. Although they failed to conquer Japan or Indonesia, they contributed to the Chinese tradition of shipbuilding.

Not only were Chinese ships the most advanced in the world but so were their techniques of navigation. In the twelfth century, sailors learned to use magnets, previously used for ceremonial and religious purposes, for navigation; an anonymous work entitled *Phing-Chou Kho T'an*, written between 1111 and 1117, explained: "The ship's pilots are acquainted with the configuration of the coasts; at night they steer by the stars, and in the daytime, by the sun. In dark weather they look at the south-pointing needle."[1] Besides knowing the direction in which they were heading, sailors needed to know where they were in relation to the land. To help them, they drew marine charts showing the outline of coasts, with headlands, rivers, and islands. At first, these charts resembled pictures of the coast as seen from a ship, but gradually, they became more accurate maps of coasts, islands, and harbors.

In 1368, a Chinese dynasty, the Ming, expelled the Mongols. At the beginning of their reign, the Ming emperors were eager to expand the influence of China in the rest of Asia. The Yongle emperor, who reigned from 1402 to 1424, was particularly interested in naval and maritime matters. In 1405, he decided to send expeditions to the countries around the Indian Ocean. To lead them, he appointed Zheng He, a Chinese Muslim mariner whose father and grandfather had made the pilgrimage to Mecca. In Zheng He's first fleet were 317 ships carrying 27,000 men. The "treasure ships" were some 400 feet long by 160 wide, the largest wooden ships ever built. They were accompanied by warships and transports for troops, horses, and drinking water. They visited Vietnam, Indonesia, and India. Later expeditions sailed to Arabia and East Africa and brought back foreign dignitaries who came to pay homage to the mighty emperor of China, as well as exotic gifts like a giraffe from the ruler of Malindi in Kenya.

The Yongle emperor died in 1424, after the sixth voyage. The first thing his successor, the Hongxi emperor, did upon ascending the throne was to order a stop to all further expeditions. Following his death nine months later, the Xuande emperor ordered one last expedition in 1432–1433 to take the foreign dignitaries back to their homes. Not long after, the government dismantled its navy. In 1500, it forbade building ships with more than two masts and, in 1525, ordered all oceangoing ships destroyed. China turned inward, leaving the oceans to foreigners.

Why this happened is still a mystery. One explanation is that the expeditions were too costly and the government needed funds for other projects. The Ming, who greatly feared the nomadic warriors of Mongolia, extended and completed the Great Wall begun 16 centuries earlier. In their new capital, Beijing, they built the Forbidden City, a great complex of palaces, buildings, and gardens. They also equipped the Grand Canal with deep-water locks and reservoirs that allowed the vital grain trade between the Yangzi Valley and northern China to flow year round without fear of pirates or storms. After that, China no longer needed oceangoing ships or costly expeditions.

Unlike China, where shipbuilding and navigation were controlled by the central government, Indian Ocean navigation was in the hands of merchants and small states. Most of the ships that plied the Indian Ocean, called *dhows*, were manned by Muslim sailors from Persia, Egypt, or Arabia. Their hulls were made of planks of Indian teak sewn together with coconut fibers and reinforced with interior ribs. This made them flexible and permitted repairs at sea, unlike ships held together with nails that had to be dragged on a beach to be repaired. Their triangular sails mounted at an angle on the mast, called *lateen*, allowed them to sail at an angle to the wind but made it difficult to change direction.

Navigation in the Indian Ocean was simpler than on any other sea. Seasonal winds called monsoons blew toward Asia in the summer and away from Asia in the winter. Navigation meant waiting for the right wind and then letting it carry you to your destination. Every round trip took a year. Sailors determined their latitude by sighting the stars with an astrolabe, an instrument first used in Persia in the eleventh century for measuring angles above the horizon. South of the equator, where sailors could not see the North Star, they calculated their latitude by measuring the height of the sun at noon, using a table of declinations

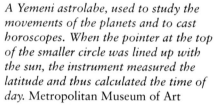

A Yemeni astrolabe, used to study the movements of the planets and to cast horoscopes. When the pointer at the top of the smaller circle was lined up with the sun, the instrument measured the latitude and thus calculated the time of day. Metropolitan Museum of Art

that gave the angle of the sun at noon for every day of the year, based on the works of the Persian astronomer Al-Khwarizi and others.

Sailors in the Mediterranean Sea faced a different set of problems than sailors faced in the Indian Ocean. The Mediterranean has fickle breezes in the warm months and storms in the winter. Since ancient times, Mediterranean shipbuilders had built galleys (oared longboats) for war and slow-sailing ships for cargo. By the fourteenth century, Venetians were building "great galleys," ships powered by both sails and oarsmen. They were large enough to carry cargo such as cloth, grain, metal goods, or spices between the Middle East and southern Europe but were not suited for long voyages outside the Mediterranean.

To carry cargo on the stormy Atlantic side of Europe, northern European shipbuilders created the cog, a round and tubby ship with a single mast carrying a large square sail. Cogs could sail only with a wind astern, and they were hard to maneuver, but they carried a lot of cargo. Their high walls made them safe from attack by ramming and boarding.

In the fifteenth century, Spanish and Portuguese shipbuilders combined the best features of galleys and cogs to produce a craft called a *caravel*, which was able to navigate any ocean in the world. Its hull had a skeleton of beams and ribs on which planks were nailed edge to edge, making it sturdy yet inexpensive. Besides the mainmast, it had a fore-mast carrying a square sail and a mizzenmast at the stern with a lateen sail. The addition of a sternpost rudder in the thirteenth century made it maneuverable in any wind. This was the type of ship used by da Gama, Columbus, and Magellan. Larger versions, the carrack and the galleon, became the mainstay of European navigation for the next four centuries.

Venturing out into the uncharted ocean was frightening to sailors who did not know how, or whether, they would return. A number of technological innovations gradually lifted the fog of uncertainty that had hampered Atlantic navigation. The first European to mention a compass was the monk Alexander Neckham in his book *De naturis rerum* in 1190. By the late thirteenth century, it was in common use in the Mediterranean and in the Middle East. In the Mediterranean, where cloudy skies had long prevented navigation during the fall and winter, ships could now sail year round without fear of losing their way. Its most dramatic impact was in northern European waters, where clouds hid the stars much of the year.

Another breakthrough in navigation was the discovery of the wind patterns over the North Atlantic Ocean. South of Lisbon on the 40th par-allel, the trade winds (so named because of their importance to merchant ships) blow from northeast to southwest, away from Europe; north of Lisbon, the westerlies blow toward the continent, bringing clouds and rain. Until the mid-fifteenth century, sailors feared that if they sailed out to sea with the trades, they would never return. The geographers who worked for the Portuguese Prince Henry the Navigator realized that the trade winds formed a *volta do mar*, or great circle of the sea, that would allow ships that sailed out with the trades to return with the westerlies. Later, when sailors had crossed the equator, they found a similar circle of winds in the South Atlantic. In the sixteenth century, explorers in the Pacific found another two voltas, one north of the equator and the other south of it. Knowing this, sailors felt confident that they could ven-ture into any body of water in the world and find their way back. This knowledge opened the doors of the Age of Exploration.

From the thirteenth century onward, Europeans drew charts ac-companied by books of sailing instructions called *portolans* (after the Italian word for "port") and constantly improved them during the years of exploration that followed. Out of sight of coasts and in uncharted waters, sailors needed a way to determine their latitude. They did so by adopting the instruments and methods long used by sailors in the Indian Ocean, such as the astrolabe and the kamal, devices used to measure the angle of the North Star with the horizon.

Before the 1780s, navigators could find their latitude, but they could not determine their longitude or position on an east–west axis. At the latitude of Europe, an error of one minute of longitude could cause a miscalculation of 20 to 30 miles—a fatal mistake that caused many shipwrecks. To solve the problem, the British Parliament in 1714 offered £20,000 (equivalent to more than $10 million in 2007) for any method of determining longitude within half a degree, or three seconds a day. In the 1730s, John Harrison, the son of a small-town carpenter and self-taught clockmaker, found ways to make clocks free of friction and immune to temperature variations. His first chronometer, H-1, was an ungainly contraption four feet high and weighing 75 pounds. Although it was accurate within one second a month, Harrison received only small subsidies to continue his experiments.

From 1737 to 1757, he built two more clocks, neither of which were tested at sea. Finally, in 1759, he presented his masterpiece, H-4, to the Board of Longitude. Tested on a ship that sailed to Jamaica and back, H-4 lost five seconds in 81 days, an accuracy far surpassing the requirements set by Parliament in 1714. However, a member of the board, the Astronomer Royal Nevil Maskelyne, had devised another method of finding longitude called lunar distances and prevented Harrison from receiving the full prize. King George III, who had long been interested in clocks, told Harrison's son William: "By God, Harrison, I will see you righted!"[2] Finally, in 1773, three years before his death at age 83, Harrison received the remainder of his prize. From then on, ships have carried chronometers inspired by Harrison's that have put marine cartography on a scientific footing and made navigation vastly safer than it had ever been.

The second technical innovation that changed the world after 1300 was gunpowder. Sometime in the ninth century, Buddhist monks from Central Asia seeking the elixir of immortality began experimenting with combinations of sulfur, charcoal, and saltpeter (potassium nitrate). Their first mixtures fizzled and smoked but did not explode. During the next four centuries, Chinese chemists concocted incendiary mixtures for use in flamethrowers, fireworks, and rocket-propelled arrows. In the thirteenth century, they learned to make an explosive mixture that could be used in bombs and grenades. To the propellant of their flamethrowers they added stones and shards of pottery, iron, or glass. This was the beginning of a new age of violence, in which the possession of firearms determined which societies had power over others.

Firearms were developed in Europe and China at about the same time. The first European to describe gunpowder was the English friar Roger Bacon in 1216. The first primitive cannon, bottle-shaped containers with a narrow opening from which a large arrow protruded, appeared in the early fourteenth century. At first, they were made of wood and hurled stone balls, but soon gunmakers learned to cast iron cannonballs and cylindrical bronze barrels. In the 1420s, gunpowder manufacturers began to compress gunpowder into pellets, a process called *corning*, that created a much more powerful explosion.

Europeans enthusiastically adopted firearms not only because their many small states and kingdoms were often at war but also because of the struggles between monarchs and the great nobles whose castles dotted the landscape. Before gunpowder, laying siege to a castle was difficult, time-consuming, and seldom successful. But large cannon, called

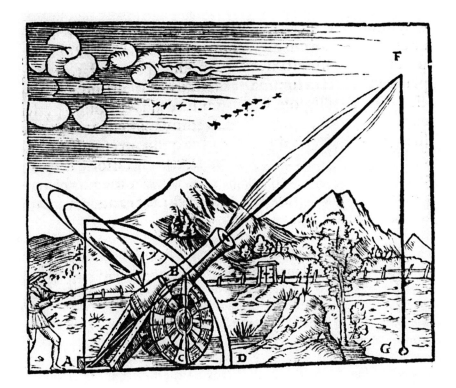

A medieval gunner uses a burning fuse at the end of a stick to ignite the powder. The illustration, from a 1561 Swiss treatise on astronomy, shows the path of the cannonball as a straight line, according to Aristotle's laws of physics, which were widely believed at the time. HIP/Art Resource, NY

bombards, could quickly breach the thickest walls. In 1446, the king of France took his cannon on a tour of the castles held by the English in Normandy and knocked them down at a rate of five a month.

The Arabs discovered gunpowder, which they called "Chinese snow," at the same time as the Europeans did. In his *Book of Fighting on Horseback and with War Engines* (1280), the Syrian military historian al-Hasan al-Rammah gave instructions for making incendiary compounds and rockets such as were used to expel the last Crusaders from Palestine in the 1290s. In the late fourteenth and fifteenth centuries, the Turks conquered the Balkans and much of the Arab world with powerful siege artillery. To lay siege to Constantinople, the most heavily defended city in Europe, they cast a huge bombard, named "Mahometta" after the prophet Muhammad, that could hurl a 600-pound stone ball. Although it cracked on the second day, other Turkish guns soon demolished the walls of the city, bringing the thousand-year-old Byzantine Empire to an end. During the sixteenth and seventeenth centuries, the Turks had the best armies in Europe, armed with the heaviest artillery. They advanced as far as the gates of Vienna in 1529 and again in 1683. During an era that we now associate with the first European empires overseas, the peoples of Europe saw themselves not as conquerors but as targets of a Turkish juggernaut.

Other rulers followed the Turkish example, setting up gun foundries and recruiting Turkish or western European gun founders to manufacture heavy artillery. In Persia, the Safavid dynasty (1501–1794) maintained itself through its possession of artillery. The Mughals, Muslim warriors from central Asia, conquered India in the sixteenth century by hiring Turkish or European gun founders and artillerymen for their campaigns against the armies of other Indian princes. Under Ivan IV "the Terrible," Russia began expanding into central Asia and across Siberia with the help of cannon and muskets. Like the Turks, all three empires succeeded because they used cannon to demolish the walls of enemy fortresses and to overawe enemy soldiers on the battlefield.

Firearms were introduced to sub-Saharan Africa by two different routes. Starting in the mid-fifteenth century, Portuguese and later other European ships brought guns and gunpowder to trade for slaves along the coast of West Africa. More disruptive was the Berber army led by the Andalusian Judar Pasha that crossed the desert in 1591 and attacked the Sudanic kingdom of Songhai with its guns. Gradually, the use of firearms spread from the coast and the western Sudan to other regions of Africa, where they were used to capture slaves as much as to conquer territory.

Warfare ensured that once states had acquired guns, they could not give them up. To this rule, there is one glaring exception: Japan. The Japanese first encountered firearms when Portuguese adventurers arrived in 1453 with two matchlocks, guns in which the powder was ignited with a match. Japanese blacksmiths quickly learned to produce such weapons in large quantities. The fifteenth and sixteenth centuries are known as the Age of the Country at War, when powerful lords battled for control of the country. At the Battle of Nagashino in 1575, an army of 38,000 men, of whom 10,000 carried guns, defeated an army of sword-wielding samurai (or Japanese knights). Japan soon had more guns than any European country.

Warfare and the proliferation of guns had serious social consequences, however. The battles showed that even a poorly trained peasant with a gun could kill a samurai, no matter how courageous, well trained, or expensively armored he might be. This threatened the position of the warrior class, who numbered half a million and were jealous of their status and their privileges, such as the right to carry swords.

In the early seventeenth century, Tokugawa Ieyasu and his descendants defeated their rivals and established a military dictatorship. In the 1630s, they began restricting the manufacture and sale of firearms. Only in two towns could gunmakers practice their trade. Civilians were forbidden to buy guns. Gradually, the government cut back its orders of firearms; by 1673, it was buying 53 large matchlocks or 334 small ones on alternate years. It also expelled all foreigners and forbade Japanese people from traveling abroad under penalty of death. For the next two centuries, no foreign power threatened Japan. The country was practically cut off from contact with the outside world and saw no reason to keep up with technological changes occurring elsewhere. Guns were forgotten until 1853, when American warships arrived in Tokyo Bay and, by firing their cannon, awoke Japan to the power of modern technology.

While central Asian rulers used heavy artillery to conquer empires, Europeans were developing cannon specifically for use on ships. The Portuguese caravels that explored the Atlantic in the early fifteenth century were equipped with small brass cannon placed on their upper decks. In the sixteenth century, cannon were placed below deck, and portholes were cut into the sides of ships to shoot out of. Instead of one big bombard, caravels carried several cannon on each side. In battles, they maneuvered alongside the enemy and fired a broadside from several cannon at once.

When Vasco da Gama's fleet arrived in India in 1498, the Muslim traders, who had long monopolized the Indian Ocean trade, realized that the Christian interlopers threatened their position. The Portuguese also understood that trading with India meant fighting the Muslims. When da Gama returned to India in 1502, he bombarded the Indian port of Calicut, destroying the local fleet. In response, the Turks sent hundreds of galleys and 15,000 men, vastly outnumbering the Portuguese. But the Turkish galleys were designed for ramming and boarding, not for gun battles at a distance, and the Portuguese quickly sank them off the coast of India. With their carracks and cannon, the Portuguese seized three strategic naval bases: Goa in India in 1510, Malacca in Malaya in 1511, and Hormuz on the Persian Gulf in 1515. Goa was the most important of the three. The Portuguese Admiral Afonso de Albuquerque wrote to King Manuel: "The taking of Goa keeps India in repose and quiet. It was folly to place all your power and strength in your navy only. . . . In ships as rotten as cork, only kept afloat by four pumps in each of them. If once Portugal should suffer a reverse at sea, your Indian possessions have not power to hold out a day longer than the Kings of this land choose to suffer it."[3] With the possession of these strong points, the Portuguese controlled the trade of the Indian Ocean, selling permits to navigate to Asian vessels and sinking those that failed to buy a permit. Although they acquired very little land, they had conquered a gunpowder empire of the sea.

States were not alone in arming their ships with cannon. Merchant ships also carried cannon for protection against pirates and corsairs, private ships with permits from one government to attack another country's ships. Given the dangers of sailing amid so many foes, the distinctions among warships, cargo ships, and pirate ships were often blurred.

The Portuguese reached Canton in southern China in 1517, announcing their arrival by firing their guns. Shortly thereafter, the scholar-official Wang-Hong wrote to the government: "The *Fo-lang-ki* [that is, Franks or Westerners] are extremely dangerous because of their artillery and their ships. . . . No weapon ever made since memorable antiquity is superior to their cannon."[4]

By then, the Chinese were familiar with cannon and other firearms but used them far less than did the Turks or Europeans because they lacked sufficient bronze. They were impressed by the Portuguese cannon, especially naval guns that could be swiveled and aimed. In the seventeenth century, high officials at the imperial court asked European Jesuit priests living in China to purchase cannon from the Portuguese. One of the Jesuits, Father Adam Schall, built and operated a gun foundry near the Imperial Palace in Beijing in the 1640s. His successor, Father

Ferdinand Verbiest, restored old Chinese bombards and manufactured 132 light cannon.

During the late seventeenth and eighteenth centuries, the Chinese conquered Tibet, Mongolia, and parts of central Asia, subduing peoples who had few or no firearms. Then, despite the Jesuits' attempts at technology transfer, Chinese interest in artillery waned because heavy cannon were useless against the horseback-riding warriors of northern and central Asia, the perennial enemy of China. As for muskets, the Chinese government feared such personal weapons falling into the hands of potential rebels more than it found them useful for defense. Unlike other Asian empires, China never fully became a gunpowder empire.

Meanwhile, European gun founders were busy advancing the technology. In contrast to China, Europe was divided into a number of competing kingdoms whose wars stirred up arms races. Portugal, the naval pioneer, was soon overshadowed by its larger neighbor, Spain, which conquered a land empire in the Americas occupied by people who had never seen firearms before. As neither country made enough firearms for their unending wars, they had to buy what they needed from the Dutch and the English. The English, who lacked the copper to make bronze, experimented with cast iron. The first iron cannon were heavy and liable to burst when fired but were much cheaper than bronze. By the late sixteenth century, the technology had improved, and England began manufacturing iron cannon in large quantities, especially for use on ships. By the early eighteenth century, England had the largest and most heavily armed fleet in the world.

The technologies of navigation and armaments that permitted the Europeans to venture out across the oceans and into distant lands had an impact far beyond geographical knowledge and military occupation. Among the most important—though often misunderstood—effects of the opening of the world was a dramatic acceleration in the transfer of plants, animals, and microorganisms from place to place.

In the history of the world, only a few dozen plants and animals have been domesticated, but the transfer of species from one place to another has been very frequent. Although we cannot date them or name the individuals responsible, wheat, rye, cattle, sheep, and goats from the Middle East, camels from Arabia, rice from Southeast Asia, horses from southern Russia, and many other plants and animals were spread throughout Eurasia and northern Africa in ancient times. Malay navigators brought the banana tree from Southeast Asia to Africa in the tenth or eleventh century. During the height of their empire from the sixth to the twelfth centuries, the Arabs brought sugar cane from India to Egypt and from there to Sicily and southern Spain. The European discovery of America triggered a great transfer of diseases, plants, and animals around the world that we call the "Columbian exchange."[5]

On his second voyage to the Caribbean, in 1493, Columbus brought seeds and cuttings of plants familiar to the Spaniards who planned to settle on the island of Hispaniola: wheat, grapevines, olive trees, sugar cane, and various fruits and vegetables. Some failed to grow in the humid tropics, but bananas, figs, melons, cabbages, and citrus fruits thrived. Later conquistadors found that the crops Spaniards prized the most—wheat for bread, olives for oil, and grapes for wine—grew well in Mexico, Peru, and Chile.

Meanwhile, the European settlers discovered the plants that the Indians cultivated: manioc or cassava in the humid tropics, maize or Indian corn in temperate regions, and potatoes in the Andes. Besides these staples, the Indians grew squash and beans, tomatoes and chili peppers, guavas and papayas, and even a plant that could be smoked: tobacco.

The Europeans who came to the New World were not just looking for land or precious metals. Their fortunes came from transporting settlers and manufactured goods from Europe and slaves from Africa to the New World in exchange for crops grown on the plantations of the Caribbean, Brazil, and southeastern North America. Tobacco and long-staple cotton were indigenous to the Americas. But Old World crops—rice, indigo, and especially sugar cane—also grew profitably in the Americas. The sugar plantations of tropical America, prison camps for their slaves, were gold mines for their owners.

In the Americas, the Europeans found ideal environments for their animals, with plentiful food and few predators. Pigs were easy to raise, for they would eat almost anything and, if not fed, could forage for themselves. Those that escaped became feral boars that ravaged their environments. Cattle—the tough Spanish longhorn breed—did poorly in the tropics but found open grasslands to their liking. Within a few decades of their introduction, vast herds of semiwild cattle populated the plains of northern Mexico, the llanos of Venezuela, and the pampas of Argentina. They provided the Spaniards with abundant meat, tallow, and hides. In the process, they overran the fields of the Indians, adding hunger to the miseries of imported diseases; as the first viceroy of Mexico wrote to the king of Spain: "May Your Lordship realize that if cattle are allowed, the Indians will be destroyed."[6]

The pride of the Spaniards was their horses. Southern Spain, from where most immigrants hailed, was a land of cattle ranches and *caballeros* (cowboys) who managed the herds. The Spaniards' horses did more to overawe the Indians than their guns, swords, or Bibles. In the mid-sixteenth century, a few horses escaped, or were stolen, from the ranches of Mexico and Chile. In the open grasslands, they thrived and multiplied. Some were captured and tamed by the Indians. Once mounted, the Indians, who had once lived precariously by hunting on foot, turned into ferocious warriors who kept the Europeans at bay for 300 years.

The transfer of Old World plants and animals to the New World was a boon for Europeans, a curse for Africans, and a mixed blessing, at best, for Indians. What the New World gave in exchange, however, was almost entirely beneficial for the peoples of the Old World.

Before the Columbian exchange, very few crops grew well in the savannas of tropical Africa and almost none in the rain forests; as a result, very few people could survive there. Sometime in the sixteenth century, the Portuguese introduced two American plants to Africa: maize, which grows well in the savannas, and manioc, a tuber that grows in almost any warm climate, even in the rain forest. Together, they provide most of the calories in the diets of Africans.

Maize also came to Europe, thriving in soils and climates where Old World cereals grew poorly. For several centuries, Europeans disdained it, believing it good only for animals. But gradually, southern Europeans developed a taste for it. The same was true of the potato in northern Europe. In the eighteenth and early nineteenth century, the English thought it fit only for the Irish, the poorest inhabitants of Great Britain, who grew potatoes almost to the exclusion of anything else. This plant, which provided more calories per acre than any other, allowed the Irish people to multiply rapidly until the 1840s, when a blight ruined the crop and reduced the population of Ireland by half. Despite this disaster, the potato now constitutes the main source of calories in the diets of northern Europeans.

The peoples of Asia also benefited from American food crops. In China and India, land that could not be turned into rice paddies proved amenable to growing maize. The same is true of manioc in Indonesia. The New World crops did not just provide more nutrients and a more varied diet, but they also occupied different ecological niches, providing a buffer in case a drought or plague damaged the traditional staple.

All in all, if we count it as a benefit for the human species to have more abundant food, then the Columbian exchange certainly increased the caloric output of the world's agriculture and, hence, the number of people it could support. It is estimated that the world's population grew from 425 million in 1500 to 545 million in 1600 (a 28 percent increase), then to 610 million in 1700 (12 percent more), and then to 900 million in 1800 (a 48 percent jump). And it is still rising. None of that would have been possible without the exchange of plants and animals between the Old World and the New.

Like gunpowder and the compass, paper originated in China and spread to the Middle East, but it had its greatest impact on Europe. For several centuries, the Chinese made paper from rattan, bamboo, or the bark of mulberry trees and used it for fans, umbrellas, and toilet paper, even for clothing and armor. A coating of starch called *sizing* made it smooth and thus useful for writing. By the Song dynasty, printing had become common, and paper mills were producing millions of sheets a year.

In 751, some Chinese paper makers, captured in a battle in central Asia, revealed their secrets to their Arab captors. Paper making reached Baghdad, then the most important city in the Arab world, in 793. In the ninth and tenth centuries, the Arab Empire was at the peak of its commercial prosperity and cultural and technological creativity, and the book trade flourished. By 900, Baghdad had more than 100 booksellers.

Europe was slow to adopt paper before the thirteenth century because fewer people were educated than in the Middle East, and their needs could be met by parchment made from sheepskins. The first paper mill in Spain dates from the early twelfth century. A hundred years later, paper manufacturing began in Italy, France, and Germany. By the fourteenth century, as more Europeans learned to read and write, the cost of parchment became exorbitant. In response to the growing demand by schools and universities, paper mills sprang up all over Europe.

Like paper, printing also originated in China. Woodblock printing, in which a page of text was carved into a block of wood, started in the eighth or ninth century and became common in the tenth. Matteo Ricci, a Jesuit priest and mathematician who lived in China from 1583 to 1610, described Chinese woodblock printing:

> Their method of making printed books is quite ingenious. The text is written in ink, with a brush made of very fine hair, on a sheet of paper which is inverted and pasted on a wooden tablet. When the paper has become thoroughly dry, its surface is scraped off quickly and with great skill, until nothing but a fine tissue bearing the characters remains on the wooden tablet. Then, with a steel engraver, the workman cuts away the surface following the outlines of the characters until these alone stand out in low relief. From such a block a skilled printer can make copies with incredible speed, turning out as many as fifteen hundred copies in a single day.[7]

In 1045, a commoner named Bi Sheng invented movable type, individual ceramic characters attached to an iron frame with wax. As China had an ideographic script in which each word was represented by a different character, setting up a print shop required a huge investment, warranted only for very large editions. Woodblock carvers could reproduce a picture of the page to be printed without knowing what it meant, but printers had to be literate to assemble the characters correctly. In the thirteenth century, Koreans began using metal type because it was well suited to their alphabetic script called *hangul*. China continued to use woodblock printing until the late nineteenth century because of the difficulty of making, storing, and retrieving the thousands of characters in Chinese writing.

One of the first uses of the new technology was printing money. The Song dynasty introduced paper money in 1024 because China did not have enough silver or copper for its growing commercial economy. Paper money allowed governments to spend far more than they received in taxes. In wartime—and the Song were often at war—such deficit spending caused runaway inflation.

Printing helped turn China into the most bureaucratic society in the world before the eighteenth century. Besides money, the Chinese government printed official decrees and handbooks of medicine, pharmacy, agriculture, mathematics, warfare, and other useful subjects; for example, *Essentials of Agriculture and Sericulture* (1273) was issued in two printings of 1,500 copies each for distribution to landowners. Commercial printers issued calendars, religious tracts, novels, sheet music, and exercise books for scholars aspiring to take the entrance examinations for government office.

Europeans were using carved wooden blocks to print cloth in the fourteenth century and to print religious images and playing cards in the fifteenth. In 1453, Johannes Gutenberg invented movable metal type independently of the Chinese. This method suited alphabetic scripts perfectly, especially after a method was found to combine illustrated woodblocks and movable type on the same page. Printing spread rapidly; by 1500, 236 towns had print shops. Twenty million books were printed before 1500 and between 140 and 200 million in the following century. Gutenberg's Bible was soon followed by Latin and Greek classics and, in the early sixteenth century, by religious tracts. Printing had a major impact on the Renaissance, the Reformation, and the scientific revolution that transformed European society.

In the Middle East, where paper had been used for centuries, printing was delayed for religious reasons. In the fifteenth century, Muslim

The Gutenberg Bible, the first large book printed with movable type, came from the Mainz print shop of Johann Gutenberg in the mid-1450s. It was printed on a wooden press, possibly modeled on a wine press, using cast metal type. The handmade paper was imported from Italy. Library of Congress LC-USZ62-51844 and LC-USZ62-87341

religious authorities banned printing on the grounds that the Qur'an, containing the sayings of the Prophet Muhammad, was originally written by hand. Not until 1706 did the Turks allow local Christians to operate a printing press, and not until the 1720s was the ban on printing in Turkish relaxed.

These technologies—ships and navigation, firearms, paper and printing—were not adopted everywhere, nor were their effects the same in all places. Instead, they favored one civilization, that of Europe, over the others. Thanks to their naval and military technologies, Europeans made their presence felt around the world from the late fifteenth century on. China, once the most innovative civilization, saw its technology fall behind that of Europe. Likewise in the Middle East, the pace of technological innovation also slowed down. How can we explain these differing rates of technological change among cultures that were in frequent contact with one another? This is one of the questions that historians find most difficult to answer.

Why did the rate of innovation in Asia slow down? The simple answer is the Mongols. These nomadic herdsmen were trained to fight on horseback from early childhood. They were also quick to adopt the most effective weapons and tactics and to recruit skilled technicians from among their settled neighbors. From 1220 to 1260, they conquered both the Middle East and China in the most rapid expansion of any empire in history. And everywhere they went, they sacked and burned cities, killed hundreds of thousands of captives, destroyed irrigation works, and ruined once-fertile land.

After the brief but destructive Mongol interlude, Arab civilization, once brilliant and innovative, became cautious and conservative. Muslim clergy resisted such innovations as mechanical clocks and printing. The Turks, who controlled the Middle East after the Mongols, kept abreast of advances in land warfare until the late eighteenth century but neglected naval shipbuilding and metallurgy and showed little interest in agricultural innovations.

Today, Westerners think of technological innovation as the key to economic growth, a rising standard of living, and national security. The Chinese had a different experience. The remarkable innovations and the flourishing economy of the Song dynasty did not protect them from the Mongols. After their defeat, China was ruled by an alien people who brought in other foreigners like Marco Polo to help them govern. After the Ming expelled the Mongols in 1368, the Chinese began to associate technological innovations and foreign ideas with defeat and humiliation.

To be sure, it was after expelling the Mongols that the Ming dynasty, a purely Chinese regime, sent out their great expeditions. But that was the personal project of the Yongle emperor, and it was quickly reversed by his successors. Because China had a single ruler whose word was law, the history of Chinese technology was particularly sensitive to the whims of whoever sat on the throne.

To administer their empire, the Ming rulers appointed mandarins, men who obtained their positions by passing examinations in classical Chinese culture. They barely tolerated military leaders and despised merchants and craftsmen, the people most interested in technology. They relied on resident Jesuits to make guns but were reluctant to permit a Chinese gun industry to develop for fear that such weapons might fall into the hands of rebels. They admired European clocks but as decorative items rather than devices to tell time. The only Western technologies they respected were astronomy, useful in making calendars, and hydraulic engineering to keep the irrigation systems functioning. In agriculture, the most essential technology of all, improvements came from adding more labor and greater skills, not from labor-saving devices. Yet until the mid-eighteenth century, China was as prosperous as any European country. Though inventions were few, familiar technologies were applied ever more skillfully, raising living standards. The Western idea that prosperity comes from technological innovation does not hold for China in this period of its history.

The Japanese experience with foreign technologies, such as firearms, differed from that of other societies, even its close neighbor China, perhaps because Japan was never occupied by the Mongols. Japanese houses, for instance, were not made to last but to be easily replaced after a fire or earthquake with renewable materials: wooden beams and floors, sliding doors and windows covered with paper, and thatched roofs. The Englishman Hugh Wilkinson, who visited Japan in the 1880s, was astonished by the construction and furnishings he found: "only a thin framework house of wood, with sliding screens of thinnest wood and paper for walls. . . . The furnishing of the place could certainly have cost but little; nor can it be said to be in accordance with our ideas of comfort. There are no chairs, or tables; neither beds, washstands, chests of drawers, basins, nor looking-glasses [that is, mirrors] encumbered the rooms."[8]

In urban sanitation, Japan was far ahead of the rest of the world until the late nineteenth century. In the early eighteenth century, the city of Edo (now Tokyo) was the largest in the world, with twice as many people as London, yet conditions there were much healthier. The *shoguns* who ruled Japan saw to it that the city had an ample supply of fresh water. Human wastes were never thrown out into gutters, streams, or cesspools, as in Europe, but collected by merchants who sold them to farmers as fertilizers. Furthermore, the Japanese were accustomed to bathing daily and washing their clothes regularly at a time when such practices were unheard of in Europe. As a result, the people suffered fewer epidemics and had a longer life expectancy than any other society.

That leaves the question of European technological creativity. Part of the reason is an accident of history. By 1241, the Mongols were about to invade western Europe when news arrived that the Great Khan Ögödei had died and all commanders had to return to Mongolia for the election of his successor. Thus did western Europe escape the barbarians who had brought ruin to the Middle East and China.

For Europeans, technological innovations provided immediate positive feedback. The Arabs and others who navigated the Indian Ocean never ventured south of Zanzibar because they did not expect to find anything there. Likewise, East Asians had little interest in exploring the Pacific because the products of Asia and Indonesia fulfilled all their needs. In contrast, the Portuguese and Spanish ventured out into the oceans seeking spices and precious metals, found them, and came back for more.

Europeans also benefited from the fact that their continent was divided into many competing states. As soon as one of them fell behind in a key technology, another quickly took its place; this is what happened in the sixteenth century when the Dutch built better ships than the Spanish and Portuguese and in the seventeenth when the English learned to make cast-iron cannon. The European states were never as authoritarian as the Asian empires, for cities and merchants always had considerable wealth and independence and competed with one another. Monarchs and their advisors learned to respect—even encourage—technological innovations as a means of enhancing their power and wealth in the competition with other nations.

The momentous events of the period from 1300 to 1800, the discoveries and contacts among all regions of the world, were both causes and consequences of key technological innovations. A surprising number of them originated in China and spread first to the Islamic world and then to Europe. In the fifteenth century, however, the technological creativity of China and Islam slowed down, while European technology was changing rapidly, driven by the competitive nature of European society and positive feedback from earlier innovations. This was especially true of naval and military technologies, those most closely connected with navigation, exploration, and trade. It was also true of plants and animals, of paper and printing, and of many other innovations.

Before 1800, the new technologies benefited a few societies—Europeans at sea and in the Americas, Turks in the Middle East and the Balkans, Mughals in India—and allowed them to grow powerful at the expense of others. But the differing rates of change among the great empires of the world widened the gap between Europeans and their rivals. After 1800, even faster technological change made the world still more lopsided.

CHAPTER 6

The First Industrial Revolution (1750–1869)

In 1765, a small group of men began meeting once a month in Birmingham, England. They called themselves the Lunar Society because they met on evenings when the moon was full, so they could find their way home afterward. Among them were the iron manufacturer Matthew Boulton, the pottery maker Josiah Wedgwood, the chemist Joseph Priestley, the scientific instrument maker James Watt, and the naturalist Erasmus Darwin. Although they left no minutes or records of their deliberations, they probably talked about commerce and manufacturing and about science and engineering. If the world we live in today is so very different from that of the eighteenth century, it is in part because men of such different backgrounds shared ideas and learned from one another and contributed to a radical change we call the Industrial Revolution.

Today, half the people on the planet, if not more, live in an industrial world. The goods we purchase, the energy we use, the way we travel, communicate, and entertain ourselves, even the food we eat are all created by large organizations using machines and processes that did not exist 200 years ago. Not even the Neolithic Revolution, with which it is often compared, transformed the human condition and the environment of the earth as completely as the Industrial Revolution.

Industrialization has four essential characteristics. The first is dividing work into a series of simple tasks carried out in factories and plantations and on construction sites. The second is using machines to replace labor and to speed up production, transportation, and communication. The third is result of the first two—namely, mass-producing goods at a lower cost than by older methods. And the fourth is generating mechanical energy from fossil fuels.

Many people have wondered why an industrial revolution did not occur in China long before the eighteenth century. For centuries, China was far ahead of the rest of the world in many important technologies.

It mass-produced iron goods, textiles, and porcelain using elaborate machines and a complex division of labor within large organizations. All it lacked was an inexpensive source of energy. Then, under the Ming dynasty, the pace of innovation slowed down. Chinese rulers and elites lost interest in technology and regarded foreign objects, such as clocks and cannon, with a mixture of curiosity and suspicion. As the population rose, labor-saving machines became less necessary.

Instead, industrialization began in Great Britain, a country that had ample deposits of coal, iron, copper, and tin and an indented coastline and navigable rivers that made transportation easy and inexpensive. By the eighteenth century, Britain was a trading nation, with a larger proportion of its inhabitants involved in commerce, shipping, and finance than any other country. Taxes and government regulations were less onerous than elsewhere. Its society encouraged innovation, entrepreneurship, and the accumulation of wealth. Businessmen were respected, and their wealth could purchase access to politics and high society. A strong patent system protected inventors. There was widespread interest in technical matters and much interaction among people in different professions, as in the Lunar Society.

Agriculture was also changing, as wealthy farmers began improving their lands, enclosing small fields into larger units, and applying new methods of preparing the soil and breeding more productive crops and animals. In the process, they forced tenant farmers and landless farm laborers to seek work in the cities, providing low-cost labor for the growing factories.

Cotton manufacturing was the most important industry of the eighteenth and early nineteenth centuries, the one that underwent the most changes and produced the greatest wealth. Before the industry arose in England, the demand for soft, colorful fabrics was met by imports of fine muslins made in India by traditional labor-intensive methods. To protect its woolen manufacturers, the British government banned the import of Indian cloth in the seventeenth and early eighteenth centuries. This enticed inventors to find ways of processing the cotton fibers in England. Their inventions turned cotton from a luxury for the rich into the economical and versatile fabric it is today.

Since the Middle Ages, European women had spun yarn on spinning wheels, an Indian or Middle Eastern invention. The foot-powered spinning wheel, introduced in the sixteenth century, increased their productivity, but it still spun only one thread. The first machine to spin several threads at once was James Hargreaves's spinning jenny (named after the daughter of another craftsman), invented in 1764. Five years later,

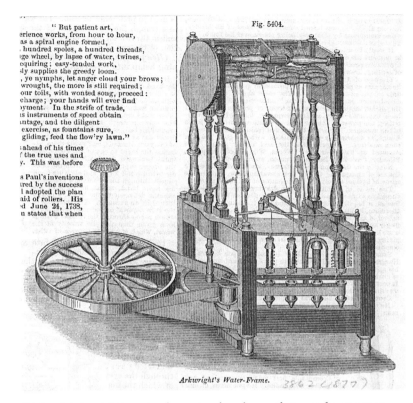

```
                                    Fig. 5404.
  " But patient art,
erience works, from hour to hour,
as a spiral engine formed,
, hundred spoles, a hundred threads,
ige wheel, by lapse of water, twines,
equiring ; easy-tended work,
sly supplies the greedy loom.
, ye nymphs, let anger cloud your brows;
wrought, the more is still required ;
our toils, with wonted song, proceed :
charge ; your hands will ever find
byment.   In the strife of trade,
is instruments of speed obtain
intage, and the diligent
   exercise, as fountains sure,
  gliding, feed the flow'ry lawn."

; ahead of his times
f the true uses and
y.  This was before

s Paul's inventions
tred by the success
l adopted the plan
aid of rollers.  His
d June 24, 1738,
n states that when
```

Arkwright's Water-Frame.

Richard Arkwright's water frame produced yarn that was far stronger
than that made by James Hargreaves's spinning jenny. Too heavy to be
moved by hand, it was powered by a waterwheel, shown here on the machine's left side. Picture Collection, The Branch Libraries, The New York
Public Library, Astor, Lenox and Tilden Foundations

the barber and entrepreneur Richard Arkwright built a more complex
machine called a water frame because it required a waterwheel to turn
it. These early machines produced an inexpensive but rough yarn. A few
years later, Samuel Crompton built a machine called a *mule* that produced a yarn that was both fine and strong. One English worker using a
mule could produce 100 times as much yarn as a worker in India.

Other machines followed in the wake of these key inventions. The
cotton gin, patented by the American inventor Eli Whitney in 1794,
separated the cotton fibers from the boll or seed of the cotton plant,
thereby making it possible to use the short-staple cotton that grew
in the southern United States. By 1861, the output of raw cotton in
the American South had grown to 1 million tons, or five-sixths of the

The cotton gin, patented by Eli Whitney in 1794, was a device that separated cotton fibers from the seeds by pulling the fibers through a screen with small wire hooks. This was especially important for the short-staple cotton grown in the American South, which had sticky seeds that were difficult to separate from the fluffy white fibers. Library of Congress LC USZ62-37836

world total. Other machines cleaned and combed the fibers. Mechanization reduced the price of yarn by 90 percent between 1790 and 1812. The enormous output of the spinning machines provided the raw material for weavers who worked in their homes at their own pace. After 1806, Edmund Cartwright's power loom began putting handloom weavers out of work. After that, mills produced both yarn and cloth.

To make cotton attractive, it had to be bleached and then printed. Cotton had traditionally been bleached by exposing it to the sun and rain for six months or by letting it soak in sour milk; neither system could keep up with the output of the new manufacturers. The solution lay in industrial chemistry, a field that the French pioneered. The cotton industry first used sulfuric acid, then sodium carbonate, and finally chlorine to bleach cloth. Once bleached, cotton could be printed much more readily than wool, linen, or silk. In the early nineteenth century, cotton manufacturers adopted copperplate printing from the book trade. Later, they introduced a method of printing long bolts of cloth with rollers engraved with a design. The result was calico, an inexpensive, bright, and colorful fabric.

One invention had consequences far beyond the textile industry. The Jacquard loom, invented by the French silk weaver Joseph-Marie Jacquard between 1801 and 1805, wove intricate patterns using silk threads of different colors. It did so by positioning a series of holes punched into cardboard cards in such a way that each hole corresponded to a particular thread to be woven, and the absence of a hole meant that thread was skipped until the next passage of the shuttle brought up the next card. In short, it was a programmed machine, the ancestor of digital computers.

The textile industry required a new system of production. Because machines were costly and needed sources of power, workers had to be concentrated in one building. By the 1790s, engineers carefully planned the design of mills and the placement of wheels, belts, shafts, and machines to maximize the efficiency of production. The socialist writer and philosopher Edward Carpenter memorably described one of these mills: "the great oblong ugly factory, in five or six tiers, all windows, alive with lights on a dark winter's morning, and again with the same lights in the evening; and all day within, the thump and scream of the machinery, and the thick smell of hot oil and cotton fluff and outside the sad smoke-laden sky, and rows of dingy streets and tall chimneys belching dirt, and the same, same outlook for miles."[1]

The textile industry produced cloth in countless colors and patterns at prices that anyone could afford. While consumers benefited, the industry had a dark side—namely, its effect on workers. Many unskilled operatives worked alongside a few skilled mechanics, supervised by foremen. Ireland, the poorest region of Britain, furnished much of the labor for the mills. Manufacturers hired women and young children to tend the machines, often making them work 12 to 16 hours a day under strict discipline, and paid them much less than they would have paid adult men. Elizabeth Bentley, a machine operator in a textile mill, described her work to a parliamentary committee in 1831:

> Explain what it is you had to do?—When the frames are full, they have to stop the frames, and take the flyers off, and take the full bobbins off, and carry them to the roller; and then put empty ones on, and set the frames on again.
>
> Does that keep you constantly on your feet?—Yes, there are so many frames and they run so quick.
>
> Your labor is very excessive?—Yes; you have not time for anything.
>
> Suppose you flagged a little, or were too late, what would they do?—Strap us.

Are they in the habit of strapping those who are last in doffing [tending machines]?—Yes.

Constantly?—Yes.

Girls as well as boys?—Yes.

Have you ever been strapped?—Yes.

Severely?—Yes.[2]

The cotton industry also revived slavery in the United States. As the mills demanded ever more raw cotton, slave owners opened up new plantations in Georgia, Alabama, and Mississippi. Although the slave trade from Africa ceased in the early nineteenth century, slave owners encouraged slaves to reproduce so that their numbers rose from 700,000 in 1790 to 3.2 million in 1850. The forced labor, racist violence, and brutal treatment to which African Americans were subjected were as much a part of the Industrial Revolution as the mills of England.

Industrialization involved much more than the production of consumer goods like cotton cloth. Two other products—iron and coal—were essential in transforming Britain into an industrial nation. Even before the Industrial Revolution, Britain had been the world's leading manufacturer and exporter of iron cannon, guns, and hardware. Iron was smelted with charcoal in small furnaces located near forests for fuel and rivers to power the bellows and hammers. Along with construction, shipbuilding, and heating, iron making contributed to deforestation, causing the price of wood to rise. In 1709, in a place called Coalbrookdale, the iron founder Abraham Darby began smelting iron with coke—that is, coal that had been baked to remove its impurities. Despite the abundance of coal and the rising price of wood, iron makers had to solve many technical problems before coke iron could compete with charcoal iron.

Most of the iron produced in the eighteenth century was cast iron, a low-cost but brittle material used to make cannon and cooking pots. In 1784, the iron worker Henry Cort built a reverberatory furnace in which the molten iron was stirred with long rods—the most strenuous job in the world—until much of the carbon burned off, leaving wrought iron, a more malleable metal less likely to crack under tension than cast iron. Thanks to these and other innovations, British iron production increased rapidly, from 25,000 tons in 1720 to 250,000 tons in 1804 and to 4 million tons in 1860, more than the rest of the world combined.

From the late eighteenth century on, the cost of iron dropped so low that it could be put to uses never dreamed of before. In 1779, Darby's grandson Abraham Darby III built an iron bridge over the River Severn. In the 1780s, architects designed factories with iron columns to support heavy vibrating machinery. By the early 1800s, engineers used iron to make beams as well as the shafts, frames, and wheels of machinery. They installed rolling mills next to iron furnaces to make flat plates that were used in mines and on construction sites. All of this required huge amounts of coal, which Britain had in abundance. Baking coal to produce coke let off flammable gases. After 1800, these gases were captured and piped to factories and streetlights to provide light that was less expensive, brighter, and safer than candles.

An important use of iron was the manufacture of cast-iron stoves, which gave off more heat while consuming far less wood or coal than fireplaces. The American Benjamin Franklin designed an improved stove—the Franklin stove—but refused to patent it, for he believed it should benefit everyone. Stoves were especially popular in cities such as London and Philadelphia, where wood was expensive.

By 1800, Britain had caught up with China in the iron and cotton industries as well as in the use of water power. But Britain's industrialization might have petered out like China's once every river had been channeled to provide power for mills and furnaces. What made industrialization an ongoing and ever-expanding revolution was the invention of machines that could extract mechanical energy from the burning of fossil fuels.

In the late seventeenth century, naturalists in Germany, France, and Britain had become aware of the pressure of the atmosphere and sought to harness its energy. Their experiments were ingenious but impractical. The first person to build a commercially successful atmospheric engine was Thomas Newcomen, an English iron and hardware maker. His first engine, completed in 1712, consisted of a boiler, a large cylinder with a piston that could travel up and down, and a rocking beam, one end of which was attached to the piston while the other was attached to a pump. When steam from the boiler entered the cylinder, the weight of the pump, pulling down its end of the rocking beam, pulled the piston up. When cold water was sprayed into the cylinder, the steam condensed, creating a vacuum, and atmospheric pressure pushed the piston down. Then the cycle was repeated.

Newcomen engines were as large as houses and burned prodigious amounts of fuel to produce only five horsepower. Because of their up-and-down motion, they could not replace waterwheels, which provided the smooth rotary motion needed to turn machinery. Yet they were the solution to the flooding of coal mines, for they were inexpensive and could pump water day and night, burning coal dust and fragments that would otherwise have gone to waste. By the end of the eighteenth century, a thousand Newcomen engines were pumping water in coal mines and towns.

The breakthrough that created the universal engine of the Industrial Revolution was the work of James Watt, a craftsman who repaired scientific instruments at the University of Glasgow in Scotland. He realized, while repairing the university's Newcomen engine, that alternately heating and cooling the cylinder wasted a great deal of fuel. The solution came to him in 1765. If the cylinder was kept hot and the steam escaped after it had done its work into a separate container that was kept constantly cold so that the steam would condense back into water, the savings in fuel would be tremendous. Watt patented his idea for a separate condenser in 1769. Actually building such an engine proved very difficult, however, for the parts had to be far more accurately made than in any previous machine of comparable size. In 1775, he formed a partnership with the engineer and fellow member of the Lunar Society Matthew Boulton. The new engines, which Boulton and Watt began manufacturing in 1776, were twice as efficient as Newcomen engines and found a market pumping water out of the tin and copper mines of Cornwall, in southern England, where coal was expensive.

Watt continued to improve his engine. He made the cylinder double acting, meaning that steam was alternately admitted to each side of the piston, doubling the power; he added a governor that automatically kept the engine running at a steady pace regardless of the load; and he invented the sun-and-planet gear that turned the back-and-forth motion of the piston into the rotary motion needed by industrial machinery.

The first customer to use a Boulton and Watt engine in a factory was Josiah Wedgwood, the pottery manufacturer and Lunar Society member. Other entrepreneurs purchased engines for breweries, flour and cotton mills, and other industrial applications. By the time their patent expired in 1799, Boulton and Watt had built almost 500 engines.

Watt knew that he could make his engine more efficient by raising the steam pressure above that of the atmosphere, but he refused to do so for fear of explosions. When his patent expired, the Englishman Richard Trevithick and the American Oliver Evans were ready to experiment with high-pressure engines. In the early 1800s, Trevithick built engines of up to 100 horsepower, twice as powerful as the biggest Watt engine. The high-pressure engines that Evans and others built in the United States did explode, many of them on Mississippi River steamers, killing their crews and passengers. Meanwhile, the British engineer Arthur Woolf realized that the steam coming out of a high-pressure cylinder still contained enough energy to push a piston in a second,

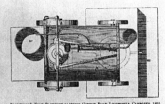

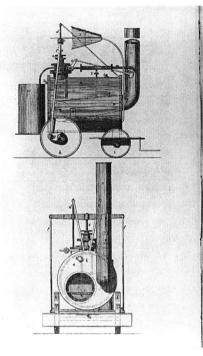

In 1802, Richard Trevithick patented the first steam locomotive to carry passengers. Trevithick's high-pressure steam engine was much more powerful for its weight than James Watt's engine. Because it did not need a water-cooled condenser as Watt's did, it could be used to propel a carriage or a locomotive on rails. Library of Congress LC-USZ62-110377

low-pressure cylinder. His compound engine produced twice as much power per pound of coal as a Watt engine. These were but the start of a long string of improvements that made the steam engine the prime mover of the Industrial Revolution well into the twentieth century. By the mid-nineteenth century, steam provided more energy than water power in Britain and later in Europe and the United States as well. The total output of steam engines in the world rose to 2 million horsepower in 1840 and to 28 million in 1880.

The steam engine was the most visible symbol of the technological progress of the early Industrial Revolution. In the nineteenth century, many Europeans and Americans equated technical progress with progress in values, or the rise of civilization as they sometimes called it. Thus, the English statesman William Huskisson boasted: "If the steam-engine be the most powerful instrument in the hands of man to alter the face of the physical world, it operates at the same time as a powerful moral lever in forwarding the great cause of civilization."[3]

Steam engines evolved in concert with three other industries: coal, iron, and machine manufacture. Coal was the fuel for engines that pumped water, turned machinery, pulled wagons, and hoisted elevators in the mines. British coal production rose from 10 million tons in 1800 to 60 million in 1850, twice as much as all the rest of the world combined. Engines and mining machinery were made of iron, and iron foundries used steam engines to produce air blasts for furnaces and to power the rolling mills that made iron sheets and rails.

By careful measurement and analysis, engineers improved the efficiency of steam engines and waterwheels. Early-eighteenth-century machines were simple enough that they could be copied by carpenters and blacksmiths. By the late eighteenth century, however, industrial equipment had to be made in machine shops. To fabricate metal parts accurately, machine designers created lathes, drills, and milling machines. By the mid-nineteenth century, such machine tools were able to turn out intricate parts for clocks, sewing machines, revolvers, rifles, and other complex devices and to make many of them interchangeable—that is, so closely alike that a new part could replace a worn one with little or no additional fitting.

Of all the machines and inventions of the Industrial Revolution, the ones that impressed contemporaries most were the applications of steam power to transportation. The first attempts to put steam engines on boats were commercial failures because they were top heavy, consumed too much coal, and produced too little power. In 1807, the American engineer Robert Fulton imported a Boulton and Watt engine and installed it on a boat popularly known as the *Clermont* on which he carried passengers on the Hudson River between New York City and Albany, New York. At first, it frightened people; Fulton's biographer Cadwallader Colden wrote that people in other boats "saw with astonishment that it was rapidly coming towards them; and when it came so near as that the noise of the machinery and paddles were heard, the crew . . . in some instances shrunk beneath their decks from the terrific sight, and left their vessels to go on shore, while others prostrated themselves, and besought Providence to protect them from the approaches of the horrible monster."[4] Soon, however, passengers were paying to take the steamer rather than ride in a carriage over bad roads, and Fulton became a respected businessman, soon followed by a host of imitators on rivers in the United States and later around the world.

Powering a riverboat was fairly simple because river water was smooth and fuel was readily available. In North America, whenever a river steamer needed fuel, the crew simply stopped and chopped down some trees along the riverbank. At sea, however, the early steam engines consumed so much coal that on long voyages there was little room left for passengers or cargo. Paddle wheels that worked well on rivers were inefficient and likely to break in ocean waves. The pounding of the engine and the heat of the firebox presented serious dangers on wooden ships. And boilers needed fresh water because salt water clogged and corroded them.

These obstacles did not faze entrepreneurs, however. As early as 1787, the foundry owner John Wilkinson built an iron barge. The first iron steamboat crossed the English Channel in 1822. Iron ships proved to be stronger and less expensive than wooden ships, and because iron was much stronger than wood, they could be built much larger. In 1838, two paddle steamers, the *Sirius* and the *Great Western*, raced each other from England to New York. Soon after, steamships with propellers were

In 1838, the Sirius, *a 703-ton British side-wheeler, was the first ship to cross the Atlantic by steam. It ran out of fuel shortly before it reached New York, but the captain refused to hoist the sails and fed spars into the furnace to complete the journey under steam power. The* Sirius *also introduced an important technical innovation, a condenser to recover the fresh water used in the boiler. I. N. Phelps Stokes Collection, Miriam and Ira D. Wallach Division of Arts, Prints and Photographs, The New York Public Library, Astor, Lenox and Tilden Foundations*

making regular voyages across the Atlantic as well as on the seas near Europe. By 1870, half the world's shipping was by steamers, although sailing ships held their own on other seas until the end of the century.

Railroads offered a bigger challenge than steamboats, but once engineers had mastered the technology, railroads proved just as important. The first person to build a steam-powered locomotive was the British engineer Richard Trevithick in 1804. His engine pulled five wagons at a speed of five miles per hour, but it was so heavy it damaged the cast-iron rails on which it ran. Not until 1820, when rolling mills began producing wrought-iron rails, was it feasible to plan an entire railroad line.

The first such line was the Stockton and Darlington Railway in northern England, opened in 1825. Four years later, the engineers George Stephenson and his son Robert built a locomotive called *Rocket* that won a contest for the fastest locomotive on the Manchester–Liverpool Railway, pulling a 20-ton train at an unheard-of 30 miles per hour. Its design was the model for all subsequent steam locomotives. Its 4-foot, 8½-inch gauge also became the standard on most other railroads in Europe and the Americas, enabling trains to switch from one company's tracks to another's.

The success of the *Rocket* and of the Manchester–Liverpool line set off a railway boom. Engineers soon developed bigger and more powerful locomotives and railway cars designed for commuter lines, local and express trains, and all kinds of freight. Locomotives and cars were only as good as the roadbed and tracks they rolled on, however. In Britain, where distances were short, engineers followed the example of canal builders and built roadbeds that were as straight and flat as possible, with gentle curves a mile or more in radius and costly stone or iron bridges. By 1850, long before other countries, railways connected all the major cities of Britain.

Safety proved to be a difficult challenge. The first casualty was William Huskisson, killed by Stephenson's *Rocket* during the Manchester–Liverpool locomotive trials. Once more railroads began operating, accidents were common. To prevent train crashes on single-track lines, engineers installed semaphores, poles with movable arms to convey signals to passing trains. They also devised the block system, by which a train could not enter a block of track until it had confirmation that no other train was present.

Americans were eager to build railroads because their country was huge and the roads were very bad. But labor was costlier and the distances much greater than in Britain, so American engineers built lines as cheaply as possible, with sharp curves and wooden trestle bridges.

To guide trains around curves and over bumpy tracks, they designed locomotives with two driving wheels at the back and a chassis with four wheels in front. They also added a tall funnel-shaped smokestack to keep burning cinders from setting fire to the adjoining forests and a triangular bumper called a *cowcatcher* to remove stray animals in a land without fences.

The first line in the United States was the Baltimore and Ohio, built in 1827–1830 using British rails and locomotives. When the U.S. government imposed heavy tariffs on imported engines, manufacturers began building the equipment in the United States. In 1840, the United States had almost 2,900 miles of track compared with 1,490 in Britain and 937 in continental Europe. By 1870, the United States had 53,000 miles of railroad track, far surpassing the 13,000 miles in Britain or Germany. The railroads open up the Midwest of the United States to European settlers and the South to planters and their slaves. They also created entire new industries such as iron foundries, rolling mills for rails, machine building, and the corporations that managed the new and very complex rail networks.

To contemporaries, one of the amazing inventions of the industrial age was a device that could transmit messages faster than a galloping horse. The first was the optical telegraph invented by Claude Chappe and adopted in 1793 by the French revolutionaries to command their armies in the field. It used towers with semaphores spaced just far enough apart to be visible from one to the next. Each position of the arms of the semaphore corresponded to a number in a code book that in turn corresponded to a word or phrase. By this system, messages could be transmitted quickly from one tower to the next. The Chappe telegraph proved so useful that it was extended throughout France. By 1850, its 556 stations covered more than 3,000 miles. Other countries built short semaphore lines for maritime or official purposes, but their capacity was too small and their cost too great to allow the public to use them.

After the Italian physicist Alessandro Volta created the first battery in 1800, scientists attempted to use electric currents to send messages. The first practical systems were developed in England and the United States. In England in 1837, Charles Wheatstone and William Cooke installed a five-wire telegraph line along the Great Western Railway to prevent accidents. When the service was opened to the public as well as railroads, it proved very popular. That same year, the American Samuel F. B. Morse patented his code, which made it possible to transmit messages by sending electrical impulses in the form of dots and dashes to

A Confederate telegraph operator stationed along the Mississippi Central Railroad taps out messages during the Civil War. The technology was crucial to military tactics on both sides of the conflict. President Lincoln visited the telegraph room at the War Department daily. Library of Congress LC-USZ61-308

represent the letters of the alphabet. He opened the first telegraph line in North America between Baltimore and Washington in 1844. Morse sent the first message on the new line to his assistant Alfred Vail: "What hath God wrought."[5] The electric telegraph was such a success that lines were quickly built in Europe and North America. By 1870, networks of lines covered both continents, and the telegraph spread to Latin America, Russia, the Ottoman Empire, and India as well.

Laying telegraph cables across the sea was far more difficult than putting up landlines. The first submarine cable connected France and Britain in 1851. During the 1850s and early 1860s, the cables laid in the Mediterranean, the Red Sea, and the Atlantic Ocean failed. In 1858, when the first cable connected Britain and the United States, President Buchanan sent a telegram to Queen Victoria: "May the Atlantic cable, under the blessing of heaven, prove to be a bond of perpetual peace and friendship between the kindred nations, and an instrument destined by Divine Providence to diffuse religion, liberty, and law throughout the world."[6] Unfortunately, the cable broke a few weeks later. Not until 1866 did cables successfully connect Europe and North America. By 1870, the technical problems were solved, and cables reached India as well. The globe was becoming wired.

Industrialization involved far more than inventions and machinery. It also changed the environment. Humans had always transformed the environments they lived in. Yet the environmental consequences of industrialization differed from all previous ones in significant ways. One cause of environmental change was the use of coal and chemicals. Coal had first been used in the Middle Ages to heat houses in London, polluting the air with foul smoke. Industrialization greatly increased the use of coal to produce iron, bricks, and glass and to power steam engines, and it did so in numerous cities and small towns. Textile industries required chemicals to clean, bleach, and dye cloth, and these chemicals, once they served their purpose, were simply dumped. Tenements for urban workers were crowded together without sanitary facilities; the inhabitants threw their garbage and sewage into the streets and ditches, from where they flowed into the nearest river. This is how a visitor to the industrial city of Manchester, England, described the River Irwell in 1862:

> the hapless river—a pretty enough stream a few miles up, with trees overhanging its banks and fringes of green sedge set thick along its edges—loses caste as it gets among the mills and print works. There are myriads of filthy things given it to wash, and whole wagonloads of poisons from dye houses and bleachyards thrown into it to carry away; steam boilers discharge into it their seething contents, and drains and sewers their fetid impurities; till at length it rolls on—here between dingy walls, there under precipices of red sand-stone—considerably less a river than a flood of liquid manure.[7]

In the early decades of industrialization, its environmental impacts were mainly felt in cities. Eventually, governments passed laws to prevent the abuses that have been described. By the end of the nineteenth century, thanks to housing codes and air- and water-pollution regulations, cities like Manchester had become more livable, even healthy.

Until 1814, industrialization was a British phenomenon, as North America and the European continent were engaged in wars and revolutions. When peace returned, industrialization spread to other countries, but it did not do so evenly. We can distinguish four groups of countries based on how industrialized they were by 1869.

In the first group were the United States and Britain's near neighbors, France, Belgium, and Prussia. Their scientists, engineers, and artisans were eager to discover the latest technological advances. Their entrepreneurs saw profit opportunities in emulating their British counterparts, and their governments shared the age-old competitiveness of European states. Foreigners flocked to Britain to learn industrial methods, to smuggle out machines or designs, and to induce British technicians and skilled workers to emigrate. Richard Arkwright's apprentice Samuel Slater emigrated to the United States in 1789 to set up a cotton mill. The American businessman Francis Cabot Lowell visited Britain in 1810–1812 and returned to New England with plans for more cotton mills. In 1799, the British entrepreneur William Cockerill moved to France to produce cloth. A few years later, he opened an iron foundry in Belgium. By 1817, he and his sons operated the largest iron foundry and textile machine manufacture in Europe. By 1830, there were 15,000 British workers in France. To catch up with Britain, the governments of western Europe encouraged workers' training and engineering education. As a result, these countries quickly caught up with British industrialization and, in some instances, surpassed it.

The nations of western Europe and North America became wealthy and powerful and eventually overcame the social and environmental problems that industrialization caused. Other parts of the world felt the effects of industrialization but found it difficult to join the club. One reason was the growth in world trade that industrialization caused. Britain exported cloth, hardware, guns, steam engines, machinery, and railroad matériel. Other countries chose to exchange their agricultural and mineral products for these manufactured goods, rather than making their own, because trading was easier, more profitable, and less disruptive to their societies than industrialization.

The most important example of this response was Russia, a country with a large population, abundant natural resources, and a government that had long been involved in European wars and diplomacy. In the eighteenth century, Russia had been the world's leading producer of iron, thanks to its forests whose lumber was used to make charcoal, but it lost that advantage when Britain began making iron with coke. In 1820, Tsar Alexander I introduced steamboat service on the Volga River. In 1830, his successor, Nicholas I, had a railroad built from Saint Petersburg to his Summer Palace in the suburbs and another one to Moscow in 1851. The government imported textile machines and iron forging equipment and offered British, German, and American engineers high salaries to come to Russia to set up cotton mills and build railroad equipment. Yet Russia remained backward. An elite of aristocrats, military officers, and government officials showed little interest in business. There was almost no ethnically Russian middle class; most of the artisans, merchants, and independent professionals were Russian Jews or Russians of German ancestry. Although the tsars and their advisors recognized the value of industry, they feared that an independent middle class and masses of urban workers would challenge their authority, as had happened in France during the revolution. They preferred to import machinery and industrial goods and pay for them through exports of grain and timber.

Latin America's failure to industrialize had different causes, but the results were similar to Russia's. Since the sixteenth century, Latin America had exported gold, silver, and agricultural products and had imported almost all manufactured goods. After they gained their independence in the early nineteenth century, the Latin American republics were dominated by an elite of landowners who were more interested in trade than in manufacturing. The wars and military coups of the mid-nineteenth century discouraged investors and retarded the industrialization of the region. Entrepreneurs, most of them immigrants from Europe, imported steam engines, textile machinery, and sugar-milling equipment at a lower cost than manufacturing them locally. Despite the growth of their economies, the Latin American republics continued to export raw materials and to depend on foreign sources for manufactured goods. Beginning in 1857, Argentina's railroads were built to transport cattle and hides from the interior to the harbor at Buenos Aires. But the money, equipment, and engineers to build and operate them came from Britain, and the official language of the railroads was not Spanish but English.

In other parts of the world, the delay resulted from a deliberate British policy to prevent industrialization. Egypt was one such case. In the early nineteenth century, the ruler of Egypt, Mohammed Ali, decided to build cotton and paper mills, sugar refineries, foundries, shipyards, and weapons factories. To protect them from the competition of lower priced British goods, he imposed high import duties. In 1839, when he went to war against the Ottoman Empire, Great Britain intervened on the Ottoman side. After defeating Egypt, Britain forced it to lower its tariffs in the name of free trade. Egypt was flooded with inexpensive British imports that put the infant Egyptian industries out of business. Mohammed Ali's successors reversed his policies and encouraged the production of wheat, cotton, and sugar for export, a policy that benefited the wealthier landowners and foreign investors but turned Egypt into an economic dependency of Europe.

India was even less able to begin industrializing because it was a colony of Britain to start with. In the eighteenth century, before the industrialization of Britain, India had been the world's leading producer of cotton cloth, all of it handmade. It also had an important shipbuilding industry thanks to abundant supplies of teak, a tropical wood that resists worms. In the early nineteenth century, British cotton exports to India, unhindered by tariffs, put Indian spinners and weavers out of work. Several Indian entrepreneurs attempted to found modern cotton mills. In 1834, Dwarkananth Tagore, a Calcutta banker, joined British businessmen to start a coal mine and a sugar refinery. In 1854, the Mumbai merchant Cowasjee Nanabhoy Dawar imported an engineer, skilled workers, and textile machines from Britain and opened a cotton mill. None of these businesses received any encouragement from the government of India.

The only industrial enterprise that really flourished in India was the railroads. The railroads of India were built by British entrepreneurs for the purpose of transporting Indian products to the harbors and British manufactured goods into the interior of India and to allow the rapid deployment of troops in the event of a rebellion. By 1870, India had the fifth longest rail network in the world, with 4,775 miles of track and 18 million passengers a year. But the engineers and equipment were all imported from Britain. As a result, this network provided few of the benefits that proved so important to the industrialization of the United States and western Europe, such as the growth of iron and machine-building industries, the employment they provided, and the profits they brought in. Despite its impressive railroad network, India remained underdeveloped well into the twentieth century.

China's first contact with a product of the Industrial Revolution brought defeat in war. By the late 1830s, the British had developed a craving for Chinese tea, but the Chinese showed little interest in foreign products, except for opium, one of India's most important exports. To stop the drug traffic, the Chinese government confiscated and destroyed the stocks of opium held by foreign merchants in Guangzhou. Outraged British merchants demanded military action to punish this blow to their business. Britain had a powerful navy, but China was a large country with few coastal cities. How could a naval power fight a large land empire?

The answer was a new machine, born in the 1830s: the steam-powered gunboat. The first gunboat to reach China was the *Nemesis*. It was 184 feet long and 29 feet wide, had an iron hull with a removable keel, and drew less than five feet of water in battle trim, so it could steam up and down shallow rivers as easily as at sea. It had two 60-horsepower steam engines driving paddle wheels and two masts and sails for ocean travel. It carried two 32-pounder cannon, 15 smaller cannon, and a rocket launcher.

Soon after the *Nemesis* arrived off the coast of China, it towed British sailing warships up the Pearl River and helped attack the Chinese forts and war junks defending Guangzhou. In 1842, five other gunboats and several steamships joined it for an attack up the Yangzi River to the Grand Canal, by which most of the north–south trade of China traveled. The commander of the expedition, Commodore Gordon Bremer, described the effect of the *Nemesis*:

> On proceeding up to Whampoa, three more dismantled forts were observed, and at four P.M. the Nemesis came into that anchorage having (in conjunction with the boats) destroyed five forts, one battery, two military stations, and nine war junks, in which were one hundred and fifteen guns and eight ginjalls [heavy muskets], thus proving to the enemy that the British flag can be displayed throughout their inner waters wherever and whenever it is thought proper by us, against any defence or mode they may adopt to prevent it.[8]

Steam and iron brought British naval power into the very heart of China and forced the Chinese government to capitulate to the long-despised "sea barbarians."

In the hundred years that followed this Opium War, three factors prevented China from industrializing: the restrictions imposed by Great Britain and other western powers, the poverty of the people, and the resistance of the traditional elite. In the eighteenth and nineteenth centuries, as the growing population of China pressed on the limited resources and arable land, poverty deepened among the farmers who constituted the majority of the people. Increasing poverty caused a decline in the cost of labor, making mechanization less attractive than it had been centuries earlier under the Song and Ming dynasties. The ruling class of mandarins educated in the Confucian classics showed little interest in commerce and viewed foreign technologies with suspicion.

The interior of Africa was relatively untouched by industrialization before 1870, protected by its disease environment and enormous transportation difficulties. But that, too, was about to change as Europeans, armed with the products of industrial manufacturers, were beginning to explore the continent and preparing to invade and conquer it.

The Industrial Revolution transformed large parts of the world within a few decades and the rest of the world shortly thereafter. Like all technological innovations, those of the Industrial Revolution gave human beings greater mastery over nature. But the changes that industrialization brought were much more rapid and disruptive than any previous era of technological change. In less than a century, people in the industrializing countries gained the ability to manufacture goods inexpensively and quickly, to travel faster and more reliably than ever before, and to communicate almost instantaneously over great distances. But industrialization also concentrated that mastery in the hands of a few, giving them power over others.

CHAPTER 7

The Acceleration of Change (1869–1939)

Around the year 1260, the English philosopher Roger Bacon predicted, "Machines may be made by which the largest ships, with only one man steering them, will be moved faster than if they were filled with rowers; wagons may be built which will move with incredible speed and without the aid of beasts; flying machines can be constructed in which a man . . . may beat the air with wings like a bird . . . machines will make it possible to go to the bottom of seas and rivers."[1]

His ideas seemed preposterous at the time and for six centuries thereafter. By the late nineteenth century, however, his predictions were beginning to come true one after another, along with many things he never dreamed of.

What caused the world to change so much, so suddenly? There were new products, certainly, and new means of mass-producing them at low cost. But standing behind these products and processes were two other more fundamental changes. One was access to cheaper, more abundant, and more useful forms of energy than humans had ever enjoyed before. The other was an exponential increase in the interactions among scientists, engineers, technicians, and businesspeople that led not only to new technologies but—for the first time—to means of creating inventions on demand.

The most noticeable transformation of the late nineteenth century was the triumph of the technologies introduced in the previous 100 years. Two events in 1869 symbolized this triumph. The first was the opening of the Suez Canal that connected the Mediterranean and the Red Seas—hence, Europe and South and East Asia. The second was the opening of the first railroad connecting the East and West Coasts of the United States.

Railroads and steamships were the most conspicuous of the technologies that transformed the world of the late nineteenth century. In 1869, a worldwide railroad boom was in full swing, as tracks spread across

Europe and North America and into Asia, Latin America, and Africa. By the early twentieth century, the industrialized countries of the world had dense rail networks that allowed passengers and freight to travel to almost every town. Ever more powerful locomotives pulled longer and heavier trains at higher speeds, requiring better signals, rails, bridges, and rolling stock. Even countries with little or no industry, such as India and Argentina, built substantial networks. In China and Africa, isolated rail lines connected mines and plantations with the nearest harbors.

Steam also conquered the oceans, as steamers replaced sailing ships on the Atlantic and Indian Oceans in the late nineteenth century and on the Pacific in the early twentieth. World trade, most of it carried by ship, increased fourfold from 1869 to 1913. Much of the growth was due to the increased size of ships; a 2,000-ton ship was considered large in the 1870s, but by the twentieth century, it would have been dwarfed by great ocean liners such as the 46,000-ton *Titanic,* which sank in 1912, or the 84,000-ton *Queen Elizabeth,* built in 1940. More important for world trade were the thousands of medium-size freighters that operated continuously, unhampered by the contrary winds that had slowed down ships in the days of sail. Ocean liners that carried passengers, mail, and costly freight followed schedules like trains, whereas tramp steamers were directed from port to port by orders telegraphed from their head-quarters in London, Paris, or New York.

The spread of railroads and of steamships brought long-distance travel within the budgets of millions of Europeans and East Asians, allowing them to emigrate to the Americas. Tying the world together was the network of submarine telegraph cables that linked Europe with North America in the 1860s, with Latin America, Asia, and Australia in the 1870s, and with Africa in the 1880s.

What made the vast extension of railroads and shipping possible was steel. Steel is a form of iron containing just enough carbon to make it hard yet flexible. Until the late nineteenth century, it had long been too rare and costly for anything but knife blades, watch springs, and other fine items. In the 1860s, William Kelly, owner of an iron manufacturing company in Kentucky, discovered that blowing air through molten pig iron turned it into steel. Improving on this process, the Englishman Henry Bessemer built converters, large containers that processed several tons of molten steel at a time at one-tenth the previous cost. Soon thereafter, another process, the Siemens-Martin open-hearth method, made steel even more economically, using scrap iron and low-grade coal. A third method, the Gilchrist-Thomas process introduced in 1875, made it possible to use phosphoric iron ores that could not be processed by

A steel ladle lifts molten steel emerging from an open-hearth furnace in U.S. Steel's Gary Steel Works, where the temperature is about 3,000 degrees Fahrenheit. Several methods of making steel, patented in the late nineteenth century, made steel cheap enough to be used for rails, bridges, ships, buildings, and even disposable cans. One method, the Siemens-Martin open-hearth furnace, used the heat of the exhaust gases to heat incoming fuel and air, thereby saving 70 to 80 percent of fuel costs. Calumet Regional Archives, Indiana University

either the Bessemer or the Siemens-Martin process. As a result, world steel output rose from 500,000 tons in 1870 to 60 million tons by 1914. Steel became inexpensive enough for the manufacture of ships, rails, bridges, and even disposable "tin" cans.

Railroads, steamships, and steel manufacturing were developed by inventors and engineers with little or no scientific training. Before the mid-nineteenth century, connections between science and technology were rare and random. Benjamin Franklin's investigation of natural electricity led him to invent the lightning rod; the research of the French chemist Claude Louis Berthollet resulted in improved methods of bleaching cloth; and the Italian scientist Alessandro Volta's experiments led him to invent the electric battery. More often, however, it was scientists who sought explanations for the phenomena that technological advances revealed; for example, steam engines had been around for a century before the French scientist Sadi Carnot formulated the laws of thermodynamics to explain how such engines transformed heat into motion. In the mid-nineteenth century, this situation began to change with the advances in electricity and chemistry.

The science of electricity was slow to produce practical results. Years went by between the invention of the wet-cell battery by the

Italian scientist Alessandro Volta in 1800 and the first electric telegraph in 1837. In the 1820s and 1830s, the Danish physicist and chemist Hans Christian Oersted and the Englishman Michael Faraday demonstrated the connections between electricity, magnetism, and motion; yet 40 years passed before their discoveries were used to generate electricity.

The telegraph industry, with its wires, batteries, and instruments, familiarized hundreds of young men with the workings of electric current and gave them tools for further experimentation. Applications other than telegraphy required electricity in far greater quantities and at much lower cost than batteries could produce, however. In the 1860s and 1870s, the German industrialist Werner von Siemens and the Belgian electrical engineer Zénobe Gramme invented dynamos that generated huge amounts of electricity. They made it possible to capture the energy of burning fuel or falling water, divide it up, and send it to distant locations where it could be used in new ways.

The first application of electricity in quantities was to power arc lights strong enough to illuminate public squares, lighthouses, and theaters. This set off a scramble to invent an electric light soft enough for homes and offices. It was well known that an electric current could turn a wire or filament white hot but only for a brief moment before it burned out. The first inventors to overcome this problem were Joseph

Edison's light bulb was a substantial improvement on earlier models, using lower current electricity, a small carbonized filament, and an improved vacuum inside the globe. This is a replica of Edison's first successful incandescent lamp, which he completed in 1879. U.S. Department of the Interior, National Park Service, Edison National Historic Site

Swan in England in 1878 and Thomas Edison in the United States in 1879. Both came up with a method of sealing a carbon filament in a vacuum so that it would shine for many hours. Swan's was the better light bulb, but Edison did more than invent a light bulb; in 1882, he built a generating plant and a system of electrical wiring, switches, fuses, and meters that lit up buildings in lower Manhattan.

Inventors and engineers soon found other uses for electricity. One of the wonders of the late nineteenth century was the telephone, patented in 1876 by an American teacher of people unable to hear and speak named Alexander Graham Bell. Bell recalled the first spoken words transmitted over a wire: "I then shouted into M [the mouthpiece] the following sentence: 'Mr. Watson—come here—I want to see you.' To my delight he came and declared that he had heard and understood what I said."[2] The instrument itself was simple; what made it truly useful was the creation of a network through which many users could call one another. This required huge investments in wires and exchanges employing dozens of switchboard operators. Starting in the early twentieth century, automatic switches allowed customers to dial a number without an operator's assistance.

On October 18, 1892, Alexander Graham Bell inaugurated the long-distance telephone line from New York to Chicago. Transcontinental telephone service was installed by 1915. Library of Congress LC-G9-Z2-28608-B

Electricity was also used to power streetcars and subways. In the 1920s, long-distance trains were electrified on heavily traveled routes. In factories, electric motors replaced steam engines. Electricity found many industrial uses, such as welding, smelting aluminum, and processing chemicals. To meet the growing demand for electricity, engineers tapped the energy of fast-flowing rivers or waterfalls with hydroelectric power plants; the first was at Niagara in upstate New York, begun in 1886, whose water turbines eventually produced 200,000 horsepower. Steam turbines, faster and more efficient than reciprocating steam engines, turned ever-larger generators. Edison's power plants had produced direct current, or DC, which is current that always flowed in the same direction; such current, however, could be transmitted only over short distances. In the 1890s, the Serbian physicist Nicola Tesla and the American industrialist George Westinghouse introduced alternating current, or AC, which could carry electric power economically at more than 100,000 volts to distant cities or factories, where transformers stepped it down to 110 or 220 volts for consumers. The electric industry soon overshadowed the mills and railroads of the first industrial revolution. By the early twentieth century, it was dominated by powerful companies like General Electric and Westinghouse in the United States and Siemens and AEG in Germany.

The industrial production of inorganic chemicals like soda and bleach began in the late eighteenth century. Organic chemicals—those containing carbon atoms—were first synthesized in 1856, when the British chemist William Perkin created a purple dye out of coal tar. Following up on his discovery, German chemists synthesized a whole series of other dyes and laid the foundation of the German chemical companies Bayer, Hoechst, and BASF, followed, in the early twentieth century, by American companies such as DuPont, Eastman Kodak, and Standard Oil. Synthetic dyes destroyed the market for natural dyes such as indigo, one of India's most important exports.

Research in chemistry created an unending stream of synthetic products. In 1866, the Swede Alfred Nobel produced dynamite, an explosive based on the dangerously unstable nitroglycerin. In 1909, the German chemist Fritz Haber discovered a method of producing ammonia, the raw material for the fertilizer and explosives industries. After World War I, chemists succeeded in creating synthetic rubber, nylon, polyethylene, and other plastics and pharmaceuticals such as aspirin, vitamins, and hormones.

Machinery and chemicals transformed agriculture from a household craft to the industrial manufacture of crops. The shift to industrial agriculture was most prominent in the midwestern region of North America, where new methods and equipment could be introduced without upsetting age-old traditions. Among the innovations were the steel moldboard plow to turn over the heavy soil of the prairie and mechanical reapers, threshers, and combines to harvest grain. In the 1890s, large farms used steam tractors and combines that could harvest up to 100 acres a day. After 1915, gasoline tractors began replacing steam engines, horses, and mules. Farmers also used increasing amounts of chemical fertilizers and pesticides. With these innovations, an American farmer in 1946 could feed 15 people, three times more than his predecessor a century before and five times more than a farmer in Asia, Africa, or southern Europe. This made food much cheaper in relation to income than in any other part of the world and allowed North American consumers to eat far more meat. Beginning in the 1890s, new machines benefited the larger, more successful farmers but put small family farms out of business, a process that is still going on.

Chemistry and electricity required much more knowledge than the craft skills that formed the basis of earlier technologies. Lone scientists had generated much new knowledge, but by the mid-nineteenth century, the problems had become so complex and the equipment needed to investigate them so costly that even wealthy amateurs could not keep up. Henceforth, new knowledge could come only from teams of researchers working in well-funded industrial or university laboratories. The skills of craftspeople and the training of engineers were still essential—and always would be—but cutting-edge technologies increasingly depended on a knowledge of science that could be acquired only in universities. The key to further advances was combining craft skills with scientific knowledge; as the English philosopher Alfred North Whitehead explained: "The greatest invention of the nineteenth century was the invention of the method of invention."[3]

The first nation to link scientific research with industry was Germany. Already by the mid-nineteenth century, that country had technical universities with research laboratories that maintained close ties to industries. From there, the idea spread to the United States with the founding of the Massachusetts Institute of Technology in 1865. In 1876, Thomas Edison established a laboratory in Menlo Park, New Jersey, that grew to include 50 workers. Although Edison was self-taught, he knew that he needed to employ physicists, chemists, and mathematicians. In this invention factory, they created the light bulb, the phonograph, and motion pictures. From Menlo Park, the idea of a research laboratory spread to electric and chemical companies nationwide. Standard Oil opened a laboratory in 1880, General Electric in 1901, DuPont in 1902, the American Telephone and Telegraph Company (AT&T) in 1911, and Eastman Kodak in 1913. By 1938, there were 2,200 corporate research laboratories in the United States alone, with the goal of developing new products and protecting their company's market share with a stream of patents. German firms also maintained extensive research programs.

Among the most important research centers was Bell Laboratories, the research arm of AT&T, which specialized in electronics. Like electricity and chemistry, electronics demonstrates the interaction of scientific theory and trial-and-error experimentation. The foundation of electronics was the theory of electromagnetic induction first proposed by Michael Faraday in 1821 and expressed mathematically by James Clerk Maxwell in 1873. In 1887, the German physics professor Heinrich Hertz demonstrated the existence of electromagnetic waves other than light. These theories and experiments soon began to produce practical applications. In 1895, the German physicist Wilhelm Roentgen discovered X-rays, an important contribution to medicine. That same year, Guglielmo Marconi discovered that electromagnetic waves could be used to transmit information. Marconi, a young man from a wealthy Irish-Italian family, experimented with homemade equipment at his family's mansion in Italy. When he realized that he could send messages in Morse code without a wire, he patented his system of "wireless telegraphy" in the hope of selling it to the British navy and merchant marine, whose ships were incommunicado as soon as they disappeared over the horizon. By 1899, he succeeded in sending a signal across the English Channel and two years later across the Atlantic Ocean.

This set off a scramble by inventors in Europe and the United States. Marconi's system used sparks to create radio waves that produced a crackling sound and could therefore be used only to transmit Morse code. Other inventors introduced devices that emitted continuous waves that could also be used to transmit voice and music. In 1915, these devices made it possible to connect New York and San Francisco by landline telephone and New York and Paris by radio telephone.

Like electricity and chemistry, electronics quickly spawned large corporations. Marconi retained control of his company, but other inventors sold or lost their patents to AT&T, RCA, and Telefunken. By the 1920s, with few exceptions, advances in electronics came out of corporate research laboratories. The age of the inventor, so dramatically illustrated by the careers of Edison and Marconi, was coming to an end. Yet their memory is kept alive in the names of major corporations: The

Marconi Company Ltd. of Britain and various Edison Electric companies in the United States.

By the 1920s, point-to-point radio linked the nations of the world and most ships at sea. In the 1930s, inexpensive shortwave radio allowed remote places and even motor vehicles to be equipped with transmitters, and individuals could afford receivers capable of tuning in the world.

Broadcasting started in 1920 with station KDKA in Pittsburgh, built by the Westinghouse Company. By the mid-1920s, radio stations were proliferating, first in the United States and then in Europe, as manufacturers built inexpensive receivers and sold them to customers eager to listen to sports, soap operas, music, and news programs. Broadcasting in the United States was subsidized by businesses that sponsored programs in exchange for the right to intersperse their programs with commercials for their products and services. By doing so, they contributed to the consumer culture that encouraged people to spend money for advertised products. In most other countries, broadcasting was a government monopoly paid for by taxes on radio receivers, and it generally broadcast culturally uplifting programs and government-approved news.

No technology has had a greater impact on human life and the environment in the twentieth century than the internal combustion engine. Unlike a steam engine in which the firebox is separated from the cylinder, in an internal combustion engine, the fuel is burned inside the cylinder, where more of its energy is converted to motion. Although the idea had been around since the seventeenth century, the first working model was built in 1859 by the French engineer Etienne Lenoir. Thousands were sold for use in factories and workshops, but they could not be used to propel vehicles because they ran on coal gas and had to be hooked up to the municipal gas mains.

The next goal was to detach the internal combustion engine from the gas mains. In 1883, the German engineer Karl Benz found a way to vaporize gasoline, a by-product of the manufacture of kerosene from petroleum. Three years later, he installed a gasoline engine on a three-wheeled cart, as did another German engineer, Gottlieb Daimler. Daimler and Benz soon joined forces and began producing the first automobiles. Other entrepreneurs built cars with steam engines or electric motors, but gasoline engines won out in the early twentieth century.

Among the many early car manufacturers, the most successful was the American machinist Henry Ford. His first car, built in 1896, was handmade like all other cars of the period. In 1908, he introduced the

Model T, a simple, sturdy, and easily maintained car designed for farmers and workers and selling for $825, a few months' salary for an average worker. Five years later, he began building cars on an assembly line, combining two well-known manufacturing techniques, interchangeable parts from the gun industry and the conveyor belt from meat-packing plants. As the belt moved the cars from worker to worker, each added one part or performed one task. Ford's assembly plant required costly machines and detailed engineering plans, but once under way, it turned out cars quickly and efficiently. In 1927, Ford was producing a car every 24 seconds and selling them for less than $300. One of the reasons they were so inexpensive is that they were all alike; as Ford once put it: "Any customer can have a car painted any color that he wants so long as it is black."[4] By then, Ford had sold 15 million Model Ts, far more than any other car before the 1950s.

Meanwhile, another kind of internal combustion engine, the diesel, competed with steam for heavier uses. Patented by the German engineer Rudolf Diesel in 1892 and manufactured from 1897 on, diesel engines were well suited for ships, heavy trucks, and industry and even began to replace steam locomotives in the 1930s.

Among the innovations that poured forth in this period of dynamic change, the one that aroused the most public excitement was the possibility of flying, a dream of humans since ancient times. In the late nineteenth century, several individuals approached the problem scientifically. Sir George Cayley in England, Otto Lilienthal in Germany, and Samuel Langley in the United States experimented with gliders and model airplanes. None of them were able to keep a human being up in the air for more than a few seconds. One solution, pioneered by the German Count Ferdinand von Zeppelin, was to build rigid lighter-than-air dirigibles that could cover huge distances at speeds up to 65 miles per hour. German dirigibles provided luxurious transatlantic passenger service until May 6, 1937, when the *Hindenburg* burst into flames in Lakehurst, New Jersey, killing 35 of the 97 people on board.

While Germans were building the first dirigibles, two American bicycle mechanics, Wilbur and Orville Wright, were studying the experiments with heavier-than-air machines and constructing a wind tunnel to test various configurations of kites, gliders, and model airplanes. They also built a gasoline engine, the first source of power that was light and strong enough to lift a human being off the ground. On December 17, 1903, their first airplane, piloted by Orville, flew for 12 seconds over a distance of 120 feet. More important, the Wright brothers were able to control their aircraft in flight. The next year, in a pasture outside

Before World War I, Zeppelins, rigid airships filled with hydrogen, were used for passenger excursions within Germany. During the war, Zeppelins dropped bombs on French and British cities. By the 1930s, regularly scheduled flights crossed the Atlantic to the United States and Brazil. Zeppelin flights ended after a fire destroyed the Hindenburg *in Lakehurst, New Jersey, in 1937.* Library of Congress LC-B2-860-1 [P&P]

Dayton, Ohio, they flew another plane over a distance of 4,080 feet in a complete circle.

After that, advances came in quick succession. By 1905, the Wright brothers were taking passengers on demonstration flights. Once news spread that humans had flown in a heavier-than-air machine, other inventors in Europe and the United States began building airplanes that soon surpassed the Wright brothers'. World War I accelerated the development of aviation. Germany, France, and Britain built some 200,000 warplanes between 1914 and 1918, mostly fighters and observation planes but a few heavy bombers as well.

After the war, air pioneers vied in breaking distance and speed records. Charles Lindbergh's and Amelia Earhart's solo nonstop flights across the Atlantic in 1927 and 1932, respectively, showed the possibilities of aviation for long-distance travel. Without military orders, aircraft manufacturers concentrated on the civilian markets, building planes to

Orville Wright pilots his 41st flight to an altitude of about 60 feet outside Dayton, Ohio, in September 1905. The Flyer III shown here was a much improved version of the original Flyer I that flew at Kitty Hawk in 1903. After 1905, the Wright brothers were involved in several patent suits and did not fly again for three years. Library of Congress LC-W861-75

carry airmail and passengers. Improvements such as retractable landing gears, metal skins, instrument flying, and reliable weather reports made aviation faster and safer. Many countries participated in the development of aviation. During the 1930s, while Britain, Germany, and Japan concentrated on warplanes in anticipation of a new conflict, the United States excelled in reliable passenger planes like the Douglas DC-3.

Industrialization involved much more than the mass production of goods such as cotton cloth, iron, and pottery and the introduction of astonishing new technologies like electricity and aviation; it also transformed the way ordinary people lived, ate, clothed, and amused themselves. In North America and western Europe, mass production required mass consumption. By the late nineteenth century, factories were turning out ready-to-wear clothing, tools, weapons, construction materials, and many other items in vast quantities at low prices. Products like the Model T were within the reach of millions of consumers. To entice people to buy their products, manufacturers placed advertisements in magazines, on billboards, and on the radio.

Industrial products changed homes and the lives of women. Factory-processed foods such as canned goods and packaged cereals made shopping and cooking easier. Pipes brought running water into the home and, with it, the "water closet," or toilet, as well as sinks and bathtubs. Gas ranges and water heaters made cooking and heating water much easier and cleaner than using wood- or coal-burning stoves. Gas was also used to light up homes at night, replacing smoky kerosene lanterns or costly candles, until gaslights were replaced in turn by electric light bulbs starting in the 1890s. Like all other household improvements, electricity first transformed the homes of the very rich, but gradually, it also became common in middle-class homes in America, Europe, and Japan. In the 1920s and 1930s, electricity was distributed to farms as well as towns. Women's lives were particularly affected by the electrification of the home. Electric washing machines reduced the drudgery of washing clothes. Electric irons eliminated the need to keep a fire in the stove to heat heavy irons, a very uncomfortable process on hot days. Vacuum cleaners made cleaning easier, and refrigerators reduced the need to shop every day.

Well-to-do families could afford all these household appliances as well as servants, but for middle-class women, the benefits were not so clear. In the nineteenth century, middle-class households had employed maids. But in the early twentieth century, as working-class women found better paying work in factories, stores, and offices, many middle-class women had to do their own housework with the help of their new appliances. Indoor kitchens, bathrooms, and toilets required cleaning. More clothing meant more washing and ironing. To keep up with increases in the quality of food and clothing and in the level of cleanliness and nutrition that middle-class standards demanded, women, especially those with large families, had to work as hard as their mothers and grandmothers did.

Although mass production benefited people in peacetime, it brought mass destruction in wartime, ruining the lives of millions of people. Nothing illustrates better the idea that power over nature gives some people power over others than the military consequences of the new technologies of the late nineteenth and early twentieth centuries.

Between 1869 and 1914, the industrial nations were at peace with one another, with the exception of two brief wars in 1870–1871 and 1904–1905. For non-Western peoples, however, this was a time of invasion and defeat. The United States, Canada, and the South American republics completed the conquest of their continents. European nations carved out large empires in Africa, Southeast Asia, and the Pacific. Russia expanded into central Asia and fought Japan for control of

Manchuria and Korea. Not since the Mongol Empire of the thirteenth century had so many lands and peoples been subjugated so fast.

What made this extraordinary expansion possible was a series of technological innovations that gave the industrial countries an overwhelming advantage over the rest of the world. Repeating rifles made of steel allowed small numbers of soldiers to defeat warriors armed with muskets, spears, or arrows. Steamships and trains could transport troops quickly to distant battlefields, while gunboats armed with machine guns or rapid-firing cannon could steam up rivers and into harbors. Quinine, a drug first extracted from the bark of the cinchona tree in 1820, protected colonial armies from malaria. In 1911, only eight years after the first human flight, the Italian army used airplanes to shoot at Libyan warriors. For those armed with the products of industry, war became a game rather than a challenge.

Then came World War I. From 1914 to 1918, the industrial nations turned their weapons on one another. To kill each other's citizens more efficiently, they devoted resources to new scientific research, accelerating the process of innovation. Heavy industries produced cannon, machine guns, and ammunition by the ton. On the battlefields, armies fired millions of cannon shells and machine-gun bullets, hoping to kill their enemies before they ran out of young men themselves. Fritz Haber, who received the Nobel Prize for his method of making fertilizer and explosives from nitrogen, also developed poison gases. Despite the heavy industry that stood behind them, soldiers on the front still had to walk across coils of barbed wire into a hail of bullets and clouds of poison, and they died by the millions.

After the mid-nineteenth century, the industrial nations and their cities had begun to take measures to alleviate the worst of the environmental problems of earlier times. In particular, they began building water and sewage systems to combat water-borne diseases, especially cholera. In the early twentieth century, the introduction of automobiles was hailed as a solution to a major urban pollution problem—namely, the tons of manure and urine that horses deposited on city streets every day.

These measures alleviated the worst aspects of local urban pollution, but environmental change accelerated and spread to entire regions. By 1890, half the world's energy came from burning coal, with a proportional increase in air pollution. Entire forests were cut down to provide lumber for buildings. In dry regions of India, Australia, and the American West, engineers planned vast irrigation projects that dwarfed the most grandiose dreams of the Pharaohs. An official of the Bureau of Reclamation in charge of public lands in the western United States expressed the ethos of this era in these words:

> The destiny of man is to possess the whole earth; and the destiny of the earth is to be subject to man. There can be no full conquest of the earth, and no real satisfaction to humanity, if large portions of the earth remain beyond his highest control. Only as parts of the earth are developed according to the best existing knowledge, and brought under human control, can man be said to possess the earth.[5]

The first industrial revolution put Great Britain ahead of its rivals in wealth and power well into the nineteenth century. Most of the technological innovations of the period 1869 to 1939, however, originated in Germany and the United States. Other countries followed closely behind, for those with sufficient resources and educated workers found the new technologies fairly easy to reproduce. Among them were not only the other nations of western Europe but also Russia and Japan.

In the mid-nineteenth century, Russia was technologically backward compared to western Europe. It had an autocratic government, a corrupt aristocracy, and a large population of peasants recently emancipated from serfdom but hardly any middle class that could imitate the technological advances of other countries. Industrialization could occur only by government decree. The first serious attempt to bring Russia into the ranks of the industrialized countries was the work of Sergei Witte, minister of finance from 1892 to 1903. Under his direction, the country began to exploit its vast reserves of coal and iron ore and build steel mills. The Russian railroad network grew from 6,670 miles in 1870 to 40,450 miles in 1915, second only to those of Canada and the United States. Most impressive was the Trans-Siberian from Moscow to the Pacific Ocean, the longest line in the world. By 1900, most of the locomotives, rails, and other equipment were made in Russia. Industrialization did not improve the living standards of the Russian people, however.

World War I and the revolutionary wars that followed until 1921 set Russian industrialization back by 20 years. Trains stopped for lack of fuel, tracks were torn up, factories closed, and scientific research came to an end. Yet the country, renamed the Union of Soviet Socialist Republics, possessed enormous reserves of raw materials and the largest population in Europe. The communists who seized power in 1917 dreamed of catching up with the West. Vladimir Lenin, their leader, announced: "Electrification plus Soviet power equals communism."

His successor after 1927, Joseph Stalin, was determined to industrialize the USSR in record time. One part of his program involved forcing peasants into collective farms served by Machine Tractor Stations that owned and operated tractors, threshers, and other agricultural machinery in imitation of the large mechanized farms of the American Midwest. This was a disaster that caused widespread famines. The other half of his program, the Five Year Plans, was successful. In the 1930s, while other industrial countries stagnated in the Great Depression, the Soviet Union tripled its output of coal, iron, steel, and heavy machinery. The government built gigantic hydroelectric dams, steel complexes, and entire industrial cities at breakneck speed. The output of the new heavy industries was enormous, but so was the cost, as consumer goods were neglected and living standards dropped below prewar levels.

The rise of Japan into the ranks of the industrial powers surprised everyone. Japan seemed an unlikely candidate for industrialization because it had no iron ore and very little coal. Instead, its technological and economic achievements came from a unique combination of government initiatives, private enterprises, and popular support.

Japan had been isolated from the rest of the world until 1854. Japanese were forbidden to travel abroad, and foreigners were not admitted into the country; only one Dutch ship a year was allowed to dock at a small island in the harbor of Nagasaki. Yet, through this tenuous source of information, educated Japanese became aware of the power and dangers of Western technology. Wrote the nationalist Sakuma Shonan: "It is astonishing that, with the invention of the steamship, the magnet, and the telegraph, they now appear to control the laws of nature."[6]

The appearance of a fleet of American warships in Tokyo Bay in 1854 set off a revolutionary change in Japanese society. In 1868, the old government was overthrown by an elite of young modernizers, the "Meiji oligarchs," who were determined to create "rich nation, strong army."[7] They knew they could do so only by transforming Japan's culture, society, and economy. To catch up with the West, the new government sent hundreds of students abroad. It also hired foreigners to teach in the Naval Academy (founded in 1870), the Imperial College of Engineering (1873), the University of Tokyo (1877), and other schools. The first professor of electrical engineering anywhere in the world was an Englishman named William Ayrton, who taught at the Imperial College of Engineering in Tokyo.

Building railways was the first priority of the new leaders starting in 1870, because they saw this industry as the key to economic development. Construction of the first line, from Tokyo to Yokohama, was directed by a British engineer. At the peak of construction, the government employed 256 foreign engineers and technicians and imported most of the rails and equipment. But the foreigners were also required to pass on their knowledge to Japanese engineers. By 1881, the foreigners had all been replaced by Japanese. By 1907, Japan no longer needed to import locomotives and other railway equipment. In building their own network with their own equipment, the Japanese obtained the products of industry, but more important, they also acquired the process and the knowledge.

Although Japan had very little coal, it had many fast-flowing rivers. Its first hydroelectric power plant was built in 1881. Electrical engineers, many of whom had studied under Ayrton or his disciples, founded such companies as Hitachi, NEC, Fujitsu, and Toshiba. To obtain the latest technology, these firms licensed patents from General Electric, AT&T, and Siemens or "reverse engineered" foreign products by taking them apart to figure out how they worked. With abundant hydroelectric power and the latest technology, Japan soon led the world in electrification. By 1935, 89 percent of Japanese households had electricity compared with 68 percent in the United States and 44 percent in Britain.

The goal of Japanese leaders was not merely to make their nation rich but also to build a strong military; in this, they were no different from the leaders of other industrial countries. From the beginning of the Meiji period, the government established arsenals and shipyards and built iron foundries and steel mills. It also placed orders with private companies such as Kawasaki and Sumitomo. As in railroads and electricity, foreign engineers and skilled workers were brought in to train Japanese and then sent home. In 1900, Japan began producing warships, some of which participated in its astonishing victory over Russia in 1905. By the 1930s, the Japanese navy was building aircraft carriers in preparation for a Pacific war. By a combination of government and private enterprise and by learning from other countries, the Japanese were able to catch up with, and in some cases surpass, the countries of western Europe.

As business competition and national rivalries kept the Western nations and Japan at the forefront of technological innovation, the rest of the world fell further behind. The failure to keep up was often due to the demands of the wealthier and more powerful advanced nations. Industrial nations needed raw materials and agricultural products that they could not produce themselves. They imported copper, tin, gold, and other metals from Asia, Africa, and South America. As their population and standard of living rose, they consumed increasing amounts of tea, coffee, sugar, bananas, and other tropical products, as well as wool and the meat and hides of cattle and sheep. To expedite this trade, they exported ships and railroad and mining equipment.

It is instructive to contrast the Japanese railways with those of India in the same period. In 1890, Japan had 1,000 miles of railroads compared to India's 17,000; 40 years later, Japan had 16,000 miles to India's 44,000. Yet, because India was a colony of Great Britain, all the engineers and almost all its rails and equipment were imported from Britain. A committee that investigated the railroads in 1921 reported:

> At the date of the last report there were employed on the railways of India about 710,000 persons; of these, roughly 700,000 were Indians and only 7,000 Europeans, a proportion of just 1 per cent. But the 7,000 were like a thin film of oil on top of a glass of water, resting upon but hardly mixing with the 700,000 below. None of the highest posts are occupied by Indians; very few even of the higher. . . .[8]

Although the Indian railroad network was far longer than the Japanese, it was an extension of British industry and technology. India received the products of industry without the benefit of learning to manufacture them.

Nonetheless, there were tentative steps toward industrialization. Tata, an Indian company, began making steel in 1911. Cotton mills were founded in many countries, as were factories to manufacture cheap consumer items such as matches, soap, and pots and pans. High technology, however, remained out of reach. Some countries were simply too poor. Others, colonies of an imperial power, were ruled by foreigners whose loyalties lay elsewhere. Yet others were held back by misguided rulers or political turmoil.

The experience of China shows how politics can delay technological change and economic development. In the mid-nineteenth century, China suffered from civil wars and foreign invasions. In the 1870s and 1880s, a group of reformers started a "self-strengthening movement" to make the country better able to defend itself, but they were stymied by the resistance of the government and its bureaucracy. In 1877, after foreigners had built a five-mile-long railroad, the government took over the line and tore it up. The contrast with the Meiji oligarchs of Japan could not be more glaring. China's time of troubles continued for another century: foreign invasions, revolutions, civil wars, and other disasters made it impossible for China to develop its industries. Only in the late twentieth century did China begin to regain its traditional place at the forefront of technology.

A person following the news of the world in the year 1939 would have learned a hard lesson about technological progress. The acceleration of change and the ever-increasing power of human beings over the forces of nature were evident everywhere: in the power to fly and to navigate underwater, to communicate instantly across oceans and continents, to feed and clothe and entertain millions of people, and to kill millions of others. This kind of progress, however, had nothing to do with progress in the ethical sense of making life better, safer, and happier. Yet it could not be stopped or even slowed down. The next stage in the history of the human race and its technologies promised to be exciting but dangerous.

CHAPTER 8

Toward a Postindustrial World (1939–2007)

On July 16, 1945, just before dawn, an atomic explosion lit the sky above New Mexico like a second sun. J. Robert Oppenheimer, leader of the scientists who had built the bomb, recalled the words of the Hindu scripture the *Bhagavad-Gita*: "I am become Death, the destroyer of worlds."[1] Three weeks later, the American bomber *Enola Gay* dropped an atomic bomb over Hiroshima, Japan. Within seconds, a fireball as powerful as several thousand tons of TNT engulfed the center of the city, turning buildings into cinders and vaporizing human beings. In outlying neighborhoods, people jumped into the river to escape the heat and were boiled alive. Those who were killed instantly—70,000 to 80,000 people—were luckier than those who survived only to die of burns and radiation poisoning in the weeks and months to come. In his memoirs, U.S. President Harry Truman recalled receiving the news: "I was greatly moved. I telephoned [Secretary of State] Byrnes aboard ship to give him the news and then said to a group of sailors around me: 'This is the greatest thing in history. It's time for us to get home.'"[2] The Hiroshima bomb and another one that destroyed Nagasaki three days later ended the deadliest war in history.

In 1914, armies and navies had gone to war with long-familiar weapons, expecting to win quickly. When victory eluded them, they turned to civilian industries to break the deadlock. The technologies introduced in World War I—explosives, poison gas, radio, and airplanes, among others—did not affect the outcome, but they taught military strategists a lesson. In the future, the armed forces could not wait for civilian inventors but would have to take an active role in developing new weapons.

When World War II began in 1939, the belligerents accelerated their military research programs to create a host of new weapons systems. Research on the frontiers of science was extremely costly, for it required the collaboration of hundreds of scientists and engineers and enormously complex equipment. Just as Edison's "invention factory" had marked a step forward in the development of technology, so did governments inaugurate the era of big science and technology in which we now live. Many of today's technologies—both military and civilian—are the result of these programs.

One such technology was radar. Since the 1920s, scientists knew that high-frequency radio waves bounced off solid objects such as mountains and ships. By measuring the delay between the transmission and reception of a radio signal, engineers could calculate the distance to the object. The military soon realized the value of such a device. By 1940, Germany and Britain had built radar networks that could detect approaching aircraft at night and in clouds. Later, American scientists at MIT's Radiation Laboratory produced high-frequency radars that could spot a submarine's periscope from a low-flying airplane and proximity fuses that made shells explode near a target rather than on impact. Research on radar cost the U.S. government the then-enormous sum of $3 billion.

Jet airplanes also appeared in World War II. The Englishman Frank Whittle had patented a gas turbine in 1931, but it was a German, Hans von Ohain, who built the first jet engine in 1939. The German Messerschmidt ME-262 jet fighter was sent into battle in 1944, too late to prevent the destruction of German cities by Allied bombers. Jet engines were more powerful and cost less to operate than piston engines. After the war, their clear superiority led to a new generation of passenger planes, beginning with the British Comet and the American Boeing 707.

Before 1940, nuclear physics was a branch of science that seemed to have few practical applications. That year, when the physicists Leo Szilard and Albert Einstein wrote to President Franklin D. Roosevelt that the Germans were trying to develop an atomic bomb, the U.S. government began to fund research in this field. In 1942, it engaged the expatriate Italian physicist Enrico Fermi to build a nuclear reactor at the University of Chicago. After that, under a secret program called the Manhattan Project, the government devoted $2 billion and the work of 43,000 people to developing atomic bombs. When Germany surrendered before the first American bombs were finished, the United States dropped them on Japan.

Computers also trace their origin to World War II. Until then, long mathematical calculations were done by hand or using simple desktop calculators. The complexity of modern weapons and the urgency of war inspired engineers and mathematicians in several countries to seek ways to hasten the process. In Germany, the electrical engineer Konrad Zuse built a digital computer using telephone switches to perform the calculations used in designing rockets. In an effort to crack the secret German military codes, the British government established a code-breaking organization where scientists led by the mathematician Max Newman built a computer called Colossus that could perform thousands of mathematical operations per second. In the United States, Howard Aiken at Harvard University built an electromechanical computer called Harvard Mark I to calculate the trajectory of artillery shells. At the University of Pennsylvania, the physicist John Mauchly and the electrical engineer J. Presper Eckert built an all-electronic machine called Electronic Numerical Integrator and Calculator, or ENIAC for short. This machine filled a large room and consumed as much electricity as a small town, but it performed calculations 1,000 times faster than electromechanical computers. For a decade after World War II, first-generation computers were so complex and costly that only the United States government could afford to operate them for military purposes.

Two staff members wire the right side of the ENIAC (Electrical Numerical Integrator and Calculator) computer developed at the University of Pennsylvania for the U.S. Army in 1943. The 30-ton, 1,800-square-foot computer, which contained more than 17,000 vacuum tubes and thousands of relays, resistors, and capacitors, ran calculations for the design of a hydrogen bomb, weather prediction, cosmic-ray studies, thermal ignition, and wind-tunnel design. U.S. Army photo

World War II also hastened the development of rocketry. The first rockets were created for the purpose of exploring outer space. In 1927, the German physicist Hermann Oberth founded a society for space travel that soon attracted a teenage enthusiast named Wernher von Braun. For more than a decade, these and other space enthusiasts wrote books and collaborated on films extolling the benefits of space travel. Meanwhile, the German army became interested in developing radical new weapons and put rocket scientists to work at a secret research facility on the Baltic Sea. In 1942, when the V-2 rocket Oberth and von Braun designed for the German army reached a height of 60 miles and a speed of 3,200 miles per hour, von Braun exclaimed: "We have invaded space with our rockets for the first time . . . we have proved rocket propulsion practicable for space travel."[3] The military used V-2s to bomb London during the last months of the war.

Another important technology developed during the war was antibiotics. In 1928, the English physician Alexander Fleming working at St. Mary's Hospital in London discovered the remarkable ability of *Penicillin notatum*, a natural mold, to kill the staphylococcus bacteria. In 1940, when the microbiologist Ernst Chain succeeded in using penicillin to cure infected mice, the British and American armies were very interested because in past wars more soldiers had died of infected wounds than in battles. By 1943, they had produced enough penicillin to protect their troops in the coming invasion of Europe.

After World War II, nations that hoped to be at the cutting edge of science and technology devoted unprecedented resources to research and development. In a report to President Roosevelt entitled *Science: The Endless Frontier* (1945), the American scientist Vannevar Bush urged his government to continue funding big scientific research projects not just for the sake of knowledge but to produce practical applications; in it, he wrote: "New products and new processes do not appear full-grown. They are founded on new principles and new conceptions, which in turn are painstakingly developed by research in the purest realms of science!"[4] His words were prescient, as the U.S. government poured ever more money into scientific research for military and medical purposes. By 1995, the U.S. government was spending $73 billion on scientific research and development, half of it devoted to advanced defense projects.

The Soviet Union, despite a smaller economy than that of the United States, spent as much on nuclear weapons as the United States. The result was a race between the United States and the Soviet Union to develop nuclear bombs and missiles and to deliver them anywhere on earth within minutes. The arms race between the two nations went on for almost 40 years. Britain joined the nuclear club in 1952, followed by France in 1960 and China in 1964.

Governments and their armed forces justified the costs and talents involved by pointing to the "spin-offs," inventions used by civilians. Radar, developed to detect enemy planes and ships in wartime, has made civilian air travel much safer and helps the police catch speeders. And Teflon, developed for use in nuclear bombs, proved useful in making nonstick frying pans.

After Hiroshima and Nagasaki, journalists said the world had entered the "atomic age." Some speculated that atomic energy would replace all other sources of energy, even powering cars and airplanes. Others feared that a third and final world war would destroy civilization. Neither has happened. Instead, nuclear physics has spawned two impressive but disappointing technologies.

Bombs were the first disappointment. In 1945, the scientists and generals in charge of the atomic bomb project were convinced that no other country could possibly match it for decades to come. They were shocked when the Soviet Union exploded its first atomic device in 1949. To counter the Soviet threat, the United States exploded a hydrogen bomb, or H-bomb, in 1952. A few months later, the Soviet Union exploded its own H-bomb. These devices were thousands of times more powerful than the bombs that fell on Japan in 1945. In a war between nations armed with such weapons, there could be no victory, only mutual suicide.

Radioactive fallout from test explosions in the atmosphere provoked a public outcry around the world, persuading the governments of nuclear-armed nations to sign a test ban treaty in 1963, a nuclear nonproliferation treaty in 1968, and strategic arms limitation treaties in the 1970s. Despite these limitations, by 1981, the nuclear powers had 50,000 nuclear warheads with a total yield of 20 billion tons of TNT, enough to destroy the planet Earth many times over. When the Soviet Union broke up in 1991, three of its successor states, Russia, Ukraine, and Kazakhstan, inherited its aging nuclear missiles. In 1998, India and Pakistan exploded nuclear devices, and Israel has the capability to quickly build its own. In 2006, North Korea claimed that it had exploded a nuclear device, but the test probably failed. Nuclear weapons are now a global phenomenon: useless in war but more dangerous than ever.

Meanwhile, scientists and engineers found another use for nuclear power. Unlike fossil-fuel plants, reactors produce heat but consume no oxygen and (when operating properly) emit no pollutants. It was obviously from the start that reactors could produce electricity inexpensively, "too cheap to meter," as one enthusiast put it. Here was the solution to the energy needs of countries that lacked oil or coal.

The first nuclear power plant began operating at Oblinsk, Russia, in 1954, followed by Calder Hall, Britain, in 1956, and Shippingport, Pennsylvania, in 1957. India and China, despite their poverty, invested in costly nuclear power plants to generate electricity and to produce plutonium with which to make atomic bombs. By 1996, 442 reactors in 32 countries were generating one-fifth of the world's electricity, with 36 more under construction. France was especially dependent on nuclear power, getting three-quarters of its electricity from nuclear plants.

Many people have become frightened of nuclear power and for good reason. In 1957, one of the first power plants, Windscale in England, leaked radioactive gases. That same year, a nuclear fuel depository at Kythtyn in Russia exploded, forcing 11,000 people to flee their homes and abandon 400,000 square miles of farmland. In 1979, a gas leak at Three Mile Island in Pennsylvania forced the evacuation of 144,000 people and cost $1 billion to clean up. Worst of all was the explosion of a nuclear reactor at Chernobyl, Ukraine, in 1986.

The Trojan Nuclear Power Plant on the Columbia River in Oregon went into service in 1975 and was immediately the target of environmental activists. After numerous technical difficulties and massive citizen efforts to shut it down, it was decommissioned in 1992. National Archives

Dozens of people were killed in the explosion and in the attempt to encase the ruined reactor in concrete, and thousands were exposed to dangerous levels of radioactivity. Radioactive fallout, carried by the wind, fell on 20 countries in Europe and made much of Ukraine and Belarus uninhabitable. The immediate cost of the disaster was estimated at $19 billion; the long-range cost is incalculable. In Japan, a leak at a nuclear fuel processing plant in 1999 exposed thousands to low-level radiation and provoked a public outrage. Construction on nuclear plants all over the world, even those nearing completion, came to a stop. Germany and Sweden decided to shut down their existing reactors. Contrary to the predictions of electricity "too cheap to meter," nuclear power proved to be the costliest technological failure in the history of the world.

During World War II, the Allies were surprised to find that Germany had developed a ballistic missile, the V-2, that was far in advance of anything they had. At the end of the war, the armies that invaded Germany seized whatever rockets and experts they could find. The U.S. Army brought Wernher von Braun to the United States, and others ended up in the Soviet Union. During the late 1940s and early 1950s, while American military research concentrated on bombs and bombers, Soviet scientists secretly worked on rockets. In October 1957, they surprised the world by launching a satellite named *Sputnik* into orbit

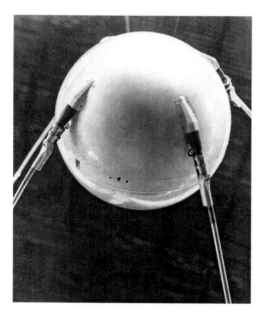

The Soviet Union launched the satellite Sputnik 1—the Russian term means "traveling companion"—on October 4, 1957, inaugurating the modern space age. Courtesy NASA History Office

around the world. It was obvious that if they could do that, their rockets could also reach any place on earth, carrying a hydrogen bomb.

The "space age," as journalists dubbed it, began as a sideshow of the arms race between the Soviet Union and the United States. While both nations were secretly building ever larger and more accurate intercontinental ballistic missiles, some with multiple atomic warheads, they engaged in a series of spectacular stunts in public. In 1961, the Soviets put the first cosmonaut, Yuri Gagarin, into orbit. Ten months later, an American team launched John Glenn into orbit. U.S. President John F. Kennedy pledged to put a man on the moon by the end of the decade, and in July 1969, two American astronauts, Neil Armstrong and Edwin "Buzz" Aldrin, landed on the moon and returned safely to Earth. The United States and the USSR also launched unmanned probes into the solar system and a space station designed to orbit Earth for years. Here was proof that with enough money and engineering talent, humans could accomplish extraordinary feats.

Luckily, rocketry found another use besides publicity stunts or missiles threatening to destroy the world. That use was satellites. In 1945, the science fiction writer Arthur C. Clarke predicted that V-2 rockets could put artificial satellites into orbit. Twelve years later, his prediction came true with *Sputnik*. It was soon followed by several Soviet and American military satellites. In 1962, *Telstar*, the first commercial communications satellite, began to transmit television broadcasts worldwide. By the 1990s, more than 100 countries had joined the International Telecommunications Satellite Organization (Intelsat), which operated satellites covering the world. The world had entered the era of instantaneous global telecommunications.

Of the thousands of satellites launched since 1957, more than half spy on other countries. The rest are used in communications, meteorology, astronomy, mapmaking, and agricultural and geological surveying. Since 1993, a system of global positioning satellites allows an inexpensive handheld device to determine one's location within a couple of yards, an unprecedented precision in navigation at sea or on land. Despite government support for the development of bombs and rockets, the most important breakthroughs of the postwar years occurred elsewhere, in electronics and in biotechnology. That is why we no longer call the times we live in the atomic age or the space age but the information age.

In 1947–1948, Walter Brattain, William Shockley, and John Bardeen, working at AT&T's Bell Laboratories, created the first transistor, a germanium crystal that could replace a vacuum tube in a radio, television, computer, or other electronic device; it was much smaller and used less electricity. In the 1960s, Texas Instruments and Fairchild Semiconductor began manufacturing transistors that were reliable and inexpensive. Electronics took another leap when scientists found a way to combine several transistors into a single integrated circuit. Once they had mastered the technique of etching transistors onto silicon wafers, or "chips," they found they could double the number of transistors—in other words, double the power of the chip—every 18 months, a formula called Moore's Law after Gordon Moore, one of the founders of the chip-manufacturing company Intel. Integrated circuits were manufactured by the billions to satisfy a huge global demand for inexpensive watches, calculators, portable radios, computers, and other consumer electronics.

In the early twentieth century, once radio became popular, inventors sought ways to transmit and receive pictures as well as sound. Producing a system that could reliably handle moving images proved to be very difficult, however, because transmitting pictures involved far more information per second than transmitting sound. Broadcasting began in a very rudimentary fashion in several countries in the late 1920s and early 1930s. All work was suspended during the war, as the equipment and expertise were needed for radar. When broadcasting resumed in the United States after the war, television quickly became as indispensable as radio had been before the war. The number of sets rose from 1 million in 1949 to 10 million two years later and 100 million in 1975; by the early twenty-first century, several billion sets were in operation worldwide.

Almost all the technologies introduced since World War II originated in the United States, Germany, Great Britain, or the Soviet Union. Yet any visit to a shopping center will show that many of the items you find there—electronic equipment, motor vehicles, appliances, cameras, and photographic supplies—are not American or European but Japanese. Japanese products began competing in the world market for consumer goods in the 1960s and quickly surpassed the West in many areas of technology, from shipbuilding to fine watches.

This came as a surprise because, at the end of the war, Japan's industries had been reduced to rubble, and electronic equipment was very hard to find. Yet there was a large demand for simple consumer goods, radios in particular. Using surplus military equipment bought on the black market or making their own parts, veterans who had been trained to build and repair radios during the war were able to assemble radios for a third to half the cost of a finished set. In 1950, they produced about three times as many receivers as the official manufacturers. Some made a living building sets to sell to others. Others opened radio repair shops. Small component manufacturers sprang up to supply the market for amateur radio hobbyists. One of these companies was Sony.

In 1946, Masaru Ibuka, an electrical engineer, and Akio Morita, an unemployed university lecturer, opened a small radio repair shop grandly titled Tokyo Telecommunications Engineering Company, or Totsuko. At the beginning, the company had no plant or equipment. Its engineers had to scrounge for surplus vacuum tubes and for rice to feed their workers and made their own tools out of pieces of metal found in the ruins of the city. In 1948, Ibuka and Morita read about the newly invented transistors in an American magazine. In 1952, on a trip to the United States, Ibuka learned that Western Electric, the manufacturing branch of the telephone giant AT&T, was willing to license the manufacture of transistors for $25,000. At the time, transistors were handmade, very costly, and used only for hearing aids. Totsuko engineers set out to build reliable high-frequency transistors that could replace vacuum tubes in radios. In 1957, they introduced the TR-63, a pocket-size, battery-operated radio that sold for $39.95 and was an instant hit. In 1960, the company began selling the first transistorized portable television. With such products and under the new name Sony, the company grew quickly. By investing in research and development, the company has remained on the cutting edge of consumer electronics technology ever since and put many American and European competitors out of business.

Other countries have wondered what the secret of Japanese commercial success is. At first, there was certainly some copying of foreign technology. But the Japanese people were technologically literate, and after the war, Japan was forbidden to rebuild its armed forces. As a result, its best engineering minds went into consumer-oriented businesses rather than into the arms industries as in the United States and the Soviet Union. Engineering attracted a larger proportion of Japanese than of any other people, and manufacturers became sensitive to the high technical demands of consumers. By the mid-1950s, Japanese consumer electronics were as advanced as any. After that, they led the world.

Radio and television are systems to broadcast information and entertainment in one direction, from the stations to the listeners or viewers. Even more useful are devices that are interactive—in other words, ones in which information can flow in both directions. Telephones are the most ubiquitous and useful interactive systems.

Telephones had been around for almost a century when they were transformed by transistors—the original goal of Bell Labs' research in the 1940s—and other advances in electronics. TAT-1, the first telephone cable across the Atlantic Ocean, was laid in 1956 and carried up to 35 simultaneous telephone conversations. The telecommunications satellites Intelsat I ("Early Bird"), launched in 1965, could handle 240 circuits. Intelsat IV, launched six years later, could carry 2,000 conversations at once. Then came fiber-optic cables, long thin strands of ultrapure glass. Lasers emitted pulses of light that could carry information without interference or blurring. This technology advanced very rapidly. The first short telephone cables in England and the United States were laid in 1983. Five years later, a consortium of AT&T, France Telecom, and British Telecom laid TAT-8, a fiber-optic cable across the Atlantic that could carry up to 40,000 simultaneous conversations. During the 1990s, fiber-optic cables linked all the nations of the world. Their capacity was so enormous—hundreds of thousands of simultaneous telephone conversations or dozens of television programs—that only a fraction was used at any one time. Long-distance communication, a costly bottleneck for centuries, was suddenly very inexpensive; the cost of a three-minute phone call from New York to London fell from $300 in 1930, to $20 in 1970, to 30¢ in 2001, and to 3¢ in 2007.

Telephone lines multiplied rapidly. By 1990, the world had 520 million telephone lines, ten times more than in 1950. Half of them were in the United States, Great Britain, and Japan; most of the others were in western Europe. Less developed countries like India and China had telephones in government offices and in the homes of important individuals. Manhattan alone had more telephones than all of Africa. Then came cellular phones, which used microwaves rather than wires. Cell phone networks spread rapidly, not only in countries with advanced wired networks but also in the major cities of poorer countries, even where there had never been a wired network. By the end of the twentieth century, the world had half a billion cell phones in addition to a billion wired phones. Futurists predicted that soon everyone on the planet would be connected to everyone else.

Overshadowing all other electronic devices was a device invented in World War II: the computer. Unlike other machines, computers are not designed to carry out a specific task but can process any sort of information, from typing papers to guiding space probes. The reason they are so versatile is that they transform all information into binary digits—strings of ones and zeros—that can be added and subtracted following the instructions contained in a program, hence the name *digital* computer. To operate, they require an input device such as a keyboard, a camera, a microphone, or a scientific instrument; a processor in the form of integrated circuits; a memory to store the program and the data going into and going out of the processor; and an output device, such as a monitor, a printer, or a device to control another machine. A major advance occurred after the war when the American mathematician John von Neumann introduced the stored program, allowing a computer to perform a multiplicity of tasks without needing to be rewired by an engineer for every job. Thereafter, improvements in computer hardware were more than matched by ingenious new software programs designed to instruct the computer to perform a specific task, from mathematical calculations to complex games.

Until the early 1950s, computers were huge and costly machines developed with the help of large government grants. The UNIVAC, built by Mauchly and Eckert at the Remington Rand Company, was used by the U.S. Census in 1951. The IBM 701, built for the U.S. Air Force to process data from its SAGE network of radar installations, propelled IBM into the dominant position in the industry, producing more than half the world's computers in 1964 and employing thousands of engineers and programmers. That year, the company introduced its System 360, a series of machines developed for businesses at a cost of $5 billion. These machines, called *mainframes,* used transistors instead of tubes and were much faster and more reliable than any previous computer. Mainframes found uses in all kinds of fields. Banks computerized their records and built networks to handle their transactions. Retail chains used them to track their inventories. Manufacturers installed robots to replace workers. Governments used computers for every job from guiding missiles to processing tax returns. Computers allowed consumers to pay with credit cards. By 1968, the United States had 70,000 computers, three-quarters of them made by IBM. IBM also dominated the world market for computers, overshadowing the efforts of German, British, and French firms. The Soviet Union, which limited its use of computers to the space program and certain key government ministries, fell further behind. IBM mainframes represented big technology, created by a powerful company with support from a government with unlimited funds. Yet the largest company in the industry in the wealthiest country on earth was shaken by competition from a completely unexpected source.

While IBM was taking over the market for large computers, integrated circuit manufacturers were designing ever more powerful chips.

In 1975, a small company named Altair discovered that a chip designed for scientific calculators could perform all the functions of a computer—albeit slowly—and offered do-it-yourself electronics buffs a computer kit for $400. In the San Francisco Bay area, a group of amateurs calling themselves the Homebrew Computer Club began sharing ideas and showing off their creations. Two of them, Stephen Wozniak and Steve Jobs, assembled their first computer circuit boards in a garage. Two years later, they launched a small desktop device called Apple II. This was no toy but a machine that could do word processing and calculate simple spreadsheets. Aimed at the small-business market, it was an instant hit and was soon selling in the tens of thousands.

IBM, waking up to the fact that it had neglected the small-business market, countered with a desktop computer of its own, the IBM personal computer, or PC, in 1981. Rather than manufacture the device, it chose to buy parts from other manufacturers, including the 8088 chip from Intel and an operating system called MS-DOS from a small software firm called Microsoft, founded by Bill Gates and Paul Allen. Since the parts, chips, and system all came from outside firms, IBM could not prevent others from manufacturing "IBM-compatible" machines. The market was soon flooded with other brands selling ever more powerful microcomputers for a price within reach of middle-class families. Meanwhile, software designers were coming out with ingenious programs that could do work, entertain, or educate people in every age group and at every level of skill. Computers, once thought of as giant "electronic brains" serviced by engineers in lab coats, had become appliances for households, businesses, and schools, and they sold in the millions.

The European countries that had fallen behind the United States in the computer revolution of the 1950s were left even further behind by the advent of personal computers. Japan, however, soon entered the personal computer market. By 1986, Japanese electronics manufacturers had overtaken American firms in designing and manufacturing certain kinds of chips. To meet the demand, they turned to suppliers in South Korea, Taiwan, Hong Kong, and Singapore, who in turn subcontracted production to factories in China. By 2001, three-quarters of the laptop computers made in the world came from Taiwan or China. Meanwhile, American firms were subcontracting software development to firms in India, whose engineers were fluent in English.

One of reasons computers became indispensable was their ability to exchange as well as process information. Computer networks, like so much else, originated with the demands of the military. In the 1960s, the U.S. Department of Defense worried that its control and communication systems were vulnerable to nuclear attack. Lawrence Roberts of the Defense Advanced Research Projects Agency developed a way to transmit messages through multiple channels by breaking them up into packets, each of which carried the addresses of the sender and the recipient and a small part of the message. Before sending a packet, special computers called *routers* checked to see if the connection was open to the next router. At the destination, a computer called a *server* reconstituted the original message. Although the parts of a message might have traveled by different routes, the message was certain to arrive complete in the end.

Packet switching made it possible to link several computers together into a network. The first computer network, named ARPANET after the Advanced Research Projects Agency, started in 1969 by linking four computers at four American universities. Protocols developed in 1983 allowed all kinds of computers to communicate with one another. The result was the Internet, a network of networks. A year later, 1,000 computers were connected to the network, mostly through telephone lines. In 1989, Tim Berners-Lee of the European Nuclear Research Center in Geneva created the World Wide Web, allowing anyone with minimal skills to access not only data in the form of words and numbers but pictures and sound as well. By 1992, a million computers were connected, and by 1998, 130 million could access the Web. Half of them were in the United States, but even China had several million networked computers. By the early twenty-first century, there were more than a billion computers in the world.

Astonishing as the advent of computers may be, it is only the beginning of the information revolution. Digitizing data means that formerly distinct information devices such as telephones, computers, television sets, game machines, cameras, printers, music players, and Internet terminals can now be combined in a process called *convergence*. Devices that can perform many of these tasks are now available and will soon be inexpensive and ubiquitous.

The information revolution has not benefited all equally, however. Nations that were once at the cutting edge have fallen behind, whereas others long considered backward, like China and India, are at the forefront of the new technology. At a time when knowledge is power as never before, wealthy countries and their wealthiest inhabitants have gained the most, while more than a third of the human race has no access to the information age and remains poor and excluded. But already, inventors are dreaming up the next devices and the next stage in the history of humankind. If recent events tell us anything, it is that we cannot predict what direction the future will take.

As dramatic as the information revolution is, future historians look-ing back at the turn of the twentieth century will probably consider an-other scientific technology even more important—namely, the biology of health and reproduction. Some new biological technologies, such as the introduction of antibiotics, were greeted with universal approval. Penicillin, introduced during World War II to save the lives of wounded soldiers, was soon followed by other antibiotics effective against vari-ous infectious diseases. As death rates dropped, people looked forward to a long and healthy life. Not until the 1990s did microbiologists no-tice the appearance of antibiotic-resistant bacteria, marking the start of an "arms race" between human ingenuity and natural evolution.

More controversial were the technologies associated with birth. Ex-periments with hormones to prevent ovulation led to the development of an oral contraceptive in 1956. Put on the market in the early 1960s, "the pill" was the most popular new pharmaceutical product in his-tory. The Catholic Church, however, opposed all artificial methods of contraception, and several predominantly Catholic countries banned its sale for several decades.

Another birth technology that gained considerable notoriety was in vitro fertilization, in which sperm and egg were united in a labora-tory, and the resulting fertilized egg then implanted in the prospective mother's uterus. The first "test-tube baby," Louise Brown, was born in 1978. Since then, there have been several thousand such births, despite religious condemnations and prohibitions in several countries.

The latest controversial technology is cloning, the creation of an animal that is the genetic "twin" of a single other animal. The sheep Dolly, born in 1997, was the first successful case of a cloned mam-mal and was followed by the cloning of other animals. As of yet, no human being has been cloned, but the technology exists. Will tech-nological determinism ("what can be done will be done") win out, or will the outcome be determined by political, moral, and scientific values?

While biomedical technologies have had important impacts on health and reproduction, biologists were probing into the very nature of life. For generations, scientists had sought to understand how living beings transferred their characteristics to their offspring. In 1953, the bacteriologist James Watson and the physicist Francis Crick, working at Cambridge University in England, came up with the double-helix model of deoxyribonucleic acid, or DNA. This substance, found in the cells of all living organisms, carried the genetic code that explained how that organism was formed and how it reproduced. Biology had now joined the information revolution, in which all information was expressed in codes.

This discovery soon began to have practical applications. In 1972–1973, two Americans, the biochemist Herbert Boyer and the geneticist Stanley Cohen, began splicing genes from one organism into another. They and others developed laboratory equipment that could replicate millions of copies of DNA in a few hours by a process called *polymerase chain reaction*. Such equipment made it possible to turn successful laboratory experiments into industrial products. Boyer founded a company named Genentech to manufacture insulin (previously obtained from pigs at great cost), thereby launching the biotechnology industry. Starting in 1984, DNA fingerprinting was able to identify perpetrators in criminal cases and even to determine the father of a child in a paternity suit. In the mid-1980s, the Human Genome Project began to decode the 100,000 genes in human DNA, at a cost of $3 billion. Access to this information promised to yield astonishing results, such as curing or preventing genetic diseases or even, possibly, determining the characteristics of an unborn child.

The genetic manipulation of human life still lies in the future, but the techniques are already common in agriculture. Humans have influenced the characteristics of useful organisms since the domestication of plants and animals 10,000 years ago. What has changed is the speed at which such changes are being introduced as a result of the new genetics.

In the 1950s, agricultural economists realized that most of the earth's arable land was already being used to grow crops and raise animals, and the human population was growing fast. The Rockefeller and Ford Foundations and the U.N. Food and Agriculture Organization funded research on improving crop yields. At the International Center for the Improvement of Corn and Wheat in Mexico, the American plant geneticist Norman Borlaug created "dwarf wheat," a variety with a short stiff straw and a large head that was resistant to disease and very responsive to inputs of fertilizer and water. Widely disseminated in Asia, it increased the harvests in India and Pakistan by 60 percent in the mid-1960s, and it benefited American, Canadian, and Australian wheat farmers even more. This success was followed by a new variety of rice, IR-8, developed at the International Rice Research Institute in the Philippines. Thanks to these and other high-yielding plants, the world's food supply was able to keep up with the growth of population for another generation.

But more was to come in the form of genetically modified crops. Research on plant DNA, started in the 1960s by large agrochemical companies, began yielding impressive results by the 1980s. They produced seeds containing DNA from several different organisms. For example, through gene splicing, DNA from bacteria could make crops resistant to frost or to insect pests. By 1998, 15 percent of the corn, 30 percent of the soybeans, and more than 50 percent of the cotton grown in the United States were genetically modified.

The benefits were immediate and obvious: less need for insecticides, fewer crop losses, and more attractive products. Yet genetically modified organisms raised fears. Would the genetic manipulation cause diseases or genetic mutations among humans in the future? Would the alien genes spread to other organisms they were never intended for, such as wild plants? Would genes designed to kill pests also wipe out butterflies and honeybees? The U.S. government argued that the new technology should be encouraged until there was proof of harm, but European governments banned genetic modification until it was proved harmless. Each side accused the other of using science as an excuse to protect its national business interests.

Fueling the onward rush of new technologies that characterize our epoch was the economic boom that has lasted from 1945 to 2007, with an interruption in the 1970s. This boom was the result of two changes, one economic and the other technological. The economic change was the reduction in tariff barriers and other impediments to trade that made it worthwhile for businesses to manufacture parts and assemble products in many parts of the world and ship them almost anywhere else. The technological advance was the container, a metal box 20 feet long by 8 feet high by 8 feet wide, that could be put on a boat, train, or truck and shipped to any destination without unpacking and repacking the contents. Starting in the 1950s, transport companies built specialized container ships and equipment to transfer containers between ships, trucks, and trains. Containers reduced the cost of shipping so dramatically that today some 90 percent of non-bulk cargo worldwide moves on container ships. The combination of lower tariffs and inexpensive freight costs has propelled the globalization of the world economy.

Every technological advance since the Industrial Revolution was designed to benefit humans—or at least some humans. Yet from the time of the first cotton mills and steam engines, it was obvious that almost all technological advances had a damaging impact on the environment. The environmental costs have become more severe as industrialization has spread and especially as humans burned ever larger quantities of fossil fuels. The problem of air pollution reached a crisis point in London in early December 1952. A cold front caused Londoners to burn more coal than usual to heat their homes, just as an inversion layer in the atmosphere trapped the smoke. The resulting pea-soup fog was so thick that traffic stopped because drivers could not see the road in front of them and movie theaters had to close because customers could not see the screen. During the four days of the fog, 4,000 people died of respiratory failure, and another 8,000 died in the days and weeks that followed. After that warning, the advanced industrial nations began to address the problem of environmental pollution and have devoted considerable resources to alleviating the worst environmental abuses of earlier times. Britain banned the burning of coal in cities, replacing it with gas, electricity, or oil. California enacted severe restrictions on auto exhaust emissions in an effort to reduce the smog in automobile-dependent Los Angeles and other urban centers. In the 1970s, the United States and European countries introduced ever more stringent pollution control and environmental protection laws. These measures were quite effective in reducing the pollution that had plagued large industrial regions like the English Midlands, southern California, and the German Ruhr. The worst environmental degradation now happened in poorer semi-industrial or nonindustrial parts of the world: deforestation in the Amazon; air pollution in Tehran, Mexico City, and Bangkok; and oil spills in the Niger Delta. Even more ominously, the ozone layer that protects Earth from ultraviolet radiation was disappearing over the Southern Hemisphere, and the eightfold increase in the consumption of petroleum since World War II was causing the entire planet to warm up, threatening to flood coastal regions and create havoc with wildlife. In short, as development spread worldwide, environmental problems went from being local to regional to global.

It is difficult to draw conclusions from events that are still happening. Looking back on almost six decades since the end of World War II, it is astonishing how many important technological developments trace their origin to that conflict. Without that war, they would certainly have been delayed. Yet many of these technologies are of questionable value either because they are too destructive, like nuclear bombs and intercontinental missiles, or involve unexpectedly high risks and costs, like nuclear power. Radar, computers, and antibiotics have certainly benefited humankind, or at least the wealthier third of the human race. Technologies developed after the war, such as transistors, color television, personal computers, the Internet, and genetically modified crops, are mostly beneficial. But like all the other technologies introduced since the Industrial Revolution, they consume energy, raw materials, and land and damage the natural world. Technology is no longer a means of survival in the face of a hostile nature but a joyride at the expense of nature. How long the joyride can continue is a question that can be answered only by the next generation.

Chronology

2.5 MILLION YEARS AGO
Oldest tools, river cobbles, used in Ethiopia

12,000–10,000 YEARS AGO
Agriculture develops in the Middle East

3RD MILLENNIUM BCE
First bronze tools used in the Middle East

2ND MILLENNIUM BCE
Malayo-Polynesians begin settling Pacific
and Indian Oceans

1700–1300 BCE
The age of chariot warfare

CA. 1500 BCE
Iron smelting in Anatolia

3RD CENTURY BCE
Romans and Chinese begin civil engineering
projects

1ST TO 8TH CENTURIES
Stirrup spreads from Afghanistan to East
Asia and Europe

1012
Champa rice introduced to China

1045
First movable type developed in China

13TH–15TH CENTURIES
Incas build road network

1327
Grand Canal links Yellow and Yangzi Rivers

14TH CENTURY
Gunpowder and cannon first used in war

1405–1433
Zheng He's treasure fleets sail to Indian
Ocean

1493
Columbian exchange of plants and animals
begins

1498
Vasco da Gama's fleet of four 90-foot-long
ships reaches India

1712
Thomas Newcomen builds first atmospheric
engine

1764
James Hargreaves invents spinning jenny

1769
James Watt patents separate condenser

1829
Rocket wins a contest and becomes the standard for steam locomotive design

1837
Charles Wheatstone and William Cooke
install the first electric telegraph line; Samuel
F. B. Morse patents his code

1838
Steamships cross the Atlantic Ocean

1866
First functioning submarine telegraph cable
begins operation across the Atlantic

1876
Alexander Graham Bell patents the
telephone

1878–1879
Thomas Edison and Joseph Swan invent incandescent electric light

1901
Guglielmo Marconi sends wireless signal
across the Atlantic Ocean

1903
Wilbur and Orville Wright fly an airplane

1908
Henry Ford introduces the Model T

1939–1945
World War II: first atom bombs, ballistic missiles, antibiotics, and computers

1957
USSR launches *Sputnik*, the first artificial satellite

1989
Tim Berners-Lee creates the Internet

1997
Dolly the cloned sheep is born

Notes

CHAPTER 2

1. *Hernán Cortés: Letters from Mexico*, ed. and trans. A. R. Pagden (New York: Grossman, 1971), 109.

2. *The Odyssey of Homer*, trans. Allen Mandelbaum, book VII. 103–111 (Berkeley: University of California Press, 1990), 136.

3. Herodotus, *The History*, trans. David Greene (Chicago: University of Chicago Press, 1987), 145.

4. Francesca Bray, *Technology and Society in Ming China (1368–1644)* (Washington, D.C.: American Historical Association, 2000), 183.

CHAPTER 3

1. Quoted in Friedrich Klemm, *A History of Western Technology*, trans. Dorothea Singer (Cambridge, Mass.: MIT Press, 1964), 51.

2. Roland A. Oliver and Brian M. Fagan, *Africa in the Iron Age, c. 500 B.C. to A.D. 1400* (Cambridge: Cambridge University Press, 1975), 69.

3. Quoted in Norman Smith, "Roman Hydraulic Technology," *Scientific American* 238 (May 1978): 155.

4. Herodotus, *The History*, book VIII, ch. 98, in *The Great Books of the Western World* (Chicago: Encyclopaedia Britannica, 1952), 6: 277.

5. Victor W. von Hagen, *The Roads That Led to Rome* (Cleveland, Ohio: World Publishing Co., 1967), 33.

6. Thucydides, *The Peloponnesian War*, book VII, ch. 23, in *The Great Books of the Western World* (Chicago: Encyclopaedia Britannica, 1952), 6: 557.

7. James Cook, *The Voyage of the* Resolution *and* Adventure, ed. John Beaglehole (Cambridge: Hakluyt Society, 1955), 154.

8. Xenophon, *Cyropaedia*, quoted in Melvin Kranzberg and Joseph Gies, *By the Sweat of Thy Brow: Work in the Western World* (New York: Putnam, 1975), 40.

9. Kax Wilson, *A History of Textiles* (Boulder, Colo.: Westview Press, 1979), 26.

CHAPTER 4

1. Marco Polo, *The Travels of Marco Polo the Venetian*, trans. and ed. William Marsden (Garden City, N.Y.: Doubleday, 1948), 223.

2. Ibid., 227.

3. Wang Zhen, *Wang Zhen nonshu* (Treatise on Agriculture), first edition 1313, 13/28b, quoted in Francesca Bray, *The Rice Economies: Technology and Development in Asian Societies* (Berkeley: University of California Press, 1986), 51.

4. *Jiaxing Prefectural Gazetteer*, 1600, quoted in Mark Elvin, *The Retreat of the Elephants: An Environmental History of China* (New Haven, Conn.: Yale University Press, 2004), 206.

5. Wang Zhen, op. cit., quoted in Mark Elvin, *The Pattern of the Chinese Past* (Stanford, Calif.: Stanford University Press, 1973), 195.

6. Hiram Bingham, *Lost City of the Incas* (New York: Atheneum, 1964), 212.

CHAPTER 5

1. Quoted in Amir D. Aczel, *The Riddle of the Compass* (New York: Harcourt, 2001), 86.

2. Quoted in Dava Sobel, *Longitude: The True Story of a Lone Genius Who Solved the Greatest Scientific Problem of His Age* (New York: Penguin, 1995), 147.

3. Quoted in G. R. Crone, *The Discovery of the East* (London: Hamish Hamilton, 1972), 54.

4. Quoted in Carlo Cipolla, *Guns, Sails, and Empires: Technological Innovation and the Early Phases of European Expansion, 1400–1700* (New York: Random House, 1965), 108.

5. Alfred Crosby, *The Columbian Exchange: Biological and Cultural Consequences of 1492* (Westport, Conn.: Greenwood Press, 1972).

6. Richard Bulliet et al., *The Earth and Its Peoples: A Global History*, 4th ed. (Boston: Houghton Mifflin, 2008), 522.

7. Matthew Ricci, *China in the Sixteenth Century*, trans. Lewis Gallagher (New York: Random House, 1943), 182.

8. Quoted in Susan B. Hanley, *Everyday Things in Premodern Japan: The Hidden Legacy of Material Culture* (Berkeley: University of California Press, 1997), 51.

CHAPTER 6

1. Edward Carpenter, *Towards Democracy*, 3rd ed. (London: T. F. Unwin, 1892), 452.

2. "Report from the Committee on the Bill to Regulate the Labour of Children in the Mills and the Factories of the United Kingdom," *British Sessional Papers*, vol. 15 (London, 1832), pp. 195–196, quoted in *The Human Record: Sources of Global History*, ed. Alfred J. Andrea and James H. Overfield, 2nd ed. (Boston: Houghton Mifflin, 1994), 263.

3. Huskisson (1824) quoted in T. K. Derry and Trevor I. Williams, *A Short History of Technology from the Earliest Times to A.D. 1900* (New York: Oxford University Press, 1961), 312.

4. Cadwallader Colden, *The Life of Robert Fulton* (New York: Kirk & Mercein, 1817), 174–175, quoted in Kirkpatrick Sale, *The Fire of His Genius: Robert Fulton and the American Dream* (New York: Free Press, 2001), 120.

5. Steven Lubar, *InfoCulture: The Smithsonian Book of Information Age Inventions* (Boston: Houghton Mifflin, 1993), 82.

6. Quoted in Paul M. Kennedy, "International Cable Communications and Strategy, 1870–1914," *English Historical Review* 86 (1971): 730.

7. Lewis Mumford, *The City in History* (New York: Harcourt, Brace & World, 1961), 459–460.

8. Letter to the Earl of Auckland, reprinted in *The London Gazette Extraordinary* (June 3, 1841): 1248.

CHAPTER 7

1. Quoted in Lynn White Jr., *Medieval Technology and Social Change* (New York: Oxford University Press, 1962), 134.

2. "American Treasures of the Library of Congress" Web site accessed June 14, 2007: http://www.loc.gov/exhibits/treasures/trr002.html.

3. Alfred North Whitehead, *Science and the Modern World* (New York: Macmillan, 1925), 98.

4. Henry Ford, *My Life and Work* (Garden City, N.Y.: Doubleday, 1922), 147.

5. John Widtsoe, *Success on Irrigation Projects* (New York: John Wiley, 1928), 138.

6. Sakuma Shonan, quoted in Thomas C. Smith, *Political Change and Industrial Development in Japan: Government Enterprise, 1868–1880* (Stanford, Calif.: Stanford University Press, 1955), 2.

7. Richard J. Samuels, *"Rich Nation, Strong Army": National Security and the Technological Transformation of Japan* (Ithaca, N.Y.: Cornell University Press, 1994).

8. India, Railway Committee, 1920–1921, *Report of the Committee Appointed by the Secretary of State for India to Enquire into the Administration and Working of the Indian Railways* (London: HMSO, 1921), 58–59.

CHAPTER 8

1. Richard Rhodes, *The Making of the Atomic Bomb* (New York: Simon & Schuster, 1988), 676.
2. Harry S. Truman, *Memoirs by Harry S. Truman*, vol. 1: *Year of Decisions* (New York: Doubleday, 1955), 421.
3. Arnold Pacey, *Technology in World Civilization* (Cambridge, Mass.: MIT Press, 1990), 175.
4. Vannevar Bush, *Science: The Endless Frontier* (Washington, D.C.: U.S. Government Printing Office, 1945), 12.

Further Reading

GLOBAL HISTORIES OF TECHNOLOGY

Carlson, Bernard, ed. *Technology in World History*. 7 vols. New York: Oxford University Press, 2005.

McClellan, James E., and Harold Dorn. *Science and Technology in World History*. Baltimore: Johns Hopkins University Press, 1999.

Pacey, Arnold. *Technology in World Civilization*. Cambridge, Mass.: MIT Press, 1990.

Smil, Vaclav. *Energy in World History*. Boulder, Colo.: Westview Press, 1994.

Technology and Culture, the quarterly journal of the Society for the History of Technology.

HISTORIES OF WESTERN TECHNOLOGY

Adams, Robert. *Paths of Fire: An Anthropologist's Inquiry into Western Civilization*. Princeton, N.J.: Princeton University Press, 1996.

Cardwell, Donald. *Wheels, Clocks, and Rockets: A History of Technology*. New York: W. W. Norton, 1995.

Kranzberg, Melvin, and Carroll W. Pursell, eds. *Technology in Western Civilization*. 2 vols. New York: Oxford University Press, 1967.

Landes, David. *Revolution in Time*. Cambridge, Mass.: Harvard University Press, 1983.

PREHISTORY

Bulliet, Richard. *Hunters, Herders, and Hamburgers: The Past and Future of Human-Animal Relationships*. New York: Columbia University Press, 2005.

Hyland, Ann. *The Horse in the Ancient World*. Westport, Conn.: Praeger, 2003.

Tattersall, Ian. *The Monkey in the Mirror: Essays on What Makes Us Human*. New York: Harcourt, 2002.

AFRICA

DeGregori, Thomas R. *Technology and the Economic Development of the African Frontier*. Cleveland, Ohio: Press of Case Western Reserve University, 1969.

Goody, Jack. *Technology, Tradition and the State in Africa*. London: Oxford University Press, 1971.

Herbert, Eugenia W. *Iron, Gender, and Power: Rituals of Transformation in African Societies*. Bloomington: Indiana University Press, 1993.

Oliver, Roland A., and Brian M. Fagan. *Africa in the Iron Age, c. 500 B.C. to A.D. 1400*. Cambridge: Cambridge University Press, 1975.

Schmidt, Peter R. *The Culture and Technology of African Iron Production*. Gainesville: University Press of Florida, 1996.

CHINA

Bray, Francesca. *The Rice Economies: Technology and Development in Asian Societies*. Berkeley: University of California Press, 1986.

Bray, Francesca. *Technology and Society in Ming China (1368–1644)*. Washington, D.C.: American Historical Association, 2000.

Elvin, Mark. *The Pattern of the Chinese Past*. Stanford, Cal: Stanford University Press, 1973.

Needham, Joseph. *The Shorter Science and Civilisation in China*. 5 vols. Cambridge: Cambridge University Press, 1978–1995.

Temple, Robert. *The Genius of China: 3,000 Years of Science, Discovery, and Invention*. 3rd ed. London: Andre Deutsch, 2007.

JAPAN

Dunn, Charles J. *Everyday Life in Traditional Japan*. Boston: Tuttle, 1969.

Hanley, Susan B. *Everyday Things in Premodern Japan: The Hidden Legacy of Material Culture*. Berkeley: University of California Press, 1997.

Low, Morris, Shigeru Nakayama, and Hitoshi Yoshioka. *Science, Technology, and Society in Contemporary Japan*. Cambridge: Cambridge University Press, 1999.

Morris-Suzuki, Tessa. *The Technological Transformation of Japan: From the 17th to the 21st Century*. Cambridge: Cambridge University Press, 1994.

Perrin, Noel. *Giving Up the Gun: Japan's Reversion to the Sword, 1543–1879*. Boulder, Colo.: Shambhala, 1979.

Samuels, Richard J. *"Rich Nation, Strong Army": National Security and the Technological Transformation of Japan*. Ithaca, N.Y.: Cornell University Press, 1994.

THE MIDDLE EAST

al-Hassan, Ahmed, and Donald Hill. *Islamic Technology: An Illustrated History*. Cambridge: Cambridge University Press, 1987.

Bloom, Jonathan M. *Paper before Print: The History and Impact of Paper in the Islamic World*. New Haven, Conn.: Yale University Press, 2001.

Bulliet, Richard. *The Camel and the Wheel*. New York: Columbia University Press, 1990.

Hill, Donald R., ed. *Studies in Medieval Islamic Technology*. Brookfield, Vt.: Ashgate-Variorum, 1998.

PREINDUSTRIAL EUROPE

Cipolla, Carlo. *Clocks and Culture, 1300–1700*. New York: W. W. Norton, 2003.

Frugoni, Chiara. *Books, Banks, Buttons: And Other Inventions from the Middle Ages*. New York: Columbia University Press, 2003.

Gimpel, Jean. *The Medieval Machine: The Industrial Revolution of the Middle Ages*. New York: Holt, Rinehart and Winston, 1976.

Kranzberg, Melvin, and Joseph Gies. *By the Sweat of Thy Brow: Work in the Western World*. New York: Putnam, 1975.

Landels, John G. *Engineering in the Ancient World*. Rev. ed. Berkeley: University of California Press, 2000.

Magnusson, Roberta J. *Water Technology in the Middle Ages: Cities, Monasteries, and Waterworks after the Roman Empire*. Baltimore: Johns Hopkins University Press, 2001.

von Hagen, Victor W. *The Roads That Led to Rome*. Cleveland, Ohio: World Publishing, 1967.

White, Lynn, Jr. *Medieval Technology and Social Change*. Oxford: Clarendon Press, 1962.

THE INDUSTRIAL REVOLUTION

Ashton, T. S. *The Industrial Revolution, 1760–1830*. New York: Oxford University Press, 1964.

Frader, Laura L. *The Industrial Revolution: A History in Documents*. New York: Oxford University Press, 2006.

Inkster, Ian, ed. *Science and Technology in History: An Approach to Industrial Development*. New Brunswick, N.J.: Rutgers University Press, 1991.

Sieferle, Rolf P. *The Subterranean Forest: Energy Systems and the Industrial Revolution*. Cambridge: White Horse Press, 2001.

Stearns, Peter. *The Industrial Revolution in World History*. Boulder, Colo.: Westview Press, 1993.

Uglow, Jenny. *The Lunar Men: Five Friends Whose Curiosity Changed the World*. New York: Farrar, Straus and Giroux, 2002.

NINETEENTH- AND TWENTIETH-CENTURY TECHNOLOGIES

Headrick, Daniel R. *The Tools of Empire: Technology and European Imperialism in the Nineteenth Century*. New York: Oxford University Press, 1981.

Levinson, Mark. *The Box: How the Shipping Container Made the World Smaller and the World Economy Bigger*. Princeton, N.J.: Princeton University Press, 2006.

Mokyr, Joel. *The Lever of Riches: Technological Creativity and Economic Progress*. New York: Oxford University Press, 1990.

Pursell, Carroll. *The Machine in America: A Social History of Technology*. Baltimore: Johns Hopkins University Press, 1995.

INFORMATION TECHNOLOGIES

Castells, Manuel. *The Internet Galaxy: Reflections on the Internet, Business, and Society*. New York: Oxford University Press, 2001.

Chandler, Alfred. *Inventing the Electronic Century: The Epic Story of the Consumer Electronics and Computer Industries*. New York: Free Press, 2001.

Eisenstein, Elizabeth. *The Printing Revolution in Early Modern Europe*. New York: Cambridge University Press, 1983.

Headrick, Daniel R. *The Invisible Weapon: Telecommunications and International Politics, 1851–1945*. New York: Oxford University Press, 1991.

Lubar, Steven. *InfoCulture: The Smithsonian Book of Information Age Inventions*. Boston: Houghton Mifflin, 1993.

SHIPS AND NAVIGATION

Aczel, Amir D. *The Riddle of the Compass: The Invention That Changed the World*. New York: Harcourt, 2001.

Casson, Lionel. *Ships and Seamanship in the Ancient World*. Baltimore: Johns Hopkins University Press, 1995.

Chaudhuri, K. N. *Trade and Civilization in the Indian Ocean: An Economic History from the Rise of Islam to 1750*. Cambridge: Cambridge University Press, 1985.

Dreyer, Edward L. *Zheng He: China and the Oceans in the Early Ming Dynasty, 1405–1433*. New York: Pearson, 2007.

Friel, Ian. *The Good Ship: Ships, Shipbuilding and Technology in England, 1200–1520*. Baltimore: Johns Hopkins University Press, 1995.

Guilmartin, John. *Gunpowder & Galleys: Changing Technology & Mediterranean Warfare at Sea in the 16th Century*. Annapolis, Md.: Naval Institute Press, 2003.

Hourani, George F. *Arab Seafaring in the Indian Ocean in Ancient and Early Medieval Times*. Revised and expanded by John Carswell. Princeton, N.J.: Princeton University Press, 1995.

Kyselka, Will. *An Ocean in Mind*. Honolulu: University of Hawaii Press, 1987.

Levathes, Louise. *When China Ruled the Seas: The Treasure Fleet of the Dragon Throne, 1405–1433*. New York: Oxford University Press, 1994.

Lewis, David. *We, the Navigators: The Ancient Art of Landfinding in the Pacific*. 2nd ed. Honolulu: University of Hawaii Press, 1994.

Parry, J. H. *The Discovery of the Sea*. New York: Dial Press, 1974.

Smith, Roger C. *Vanguard of Empire: Ships of Exploration in the Age of Columbus*. New York: Oxford University Press, 1993.

Sobel, Dava. *Longitude: The True Story of a Lone Genius Who Solved the Greatest Scientific Problem of His Time*. New York: Penguin, 1995.

TEXTILE TECHNOLOGY

Barber, Elizabeth W. *Women's Work: The First 20,000 Years: Women, Cloth, and Society in Early Times*. New York: W. W. Norton, 1994.

Bray, Francesca. *Technology and Gender: Fabrics of Power in Late Imperial China*. Berkeley: University of California Press, 1997.

Wilson, Kax. *A History of Textiles*. Boulder, Colo.: Westview Press, 1979.

WEAPONS AND MILITARY TECHNOLOGY

Bishop, M. C., and J. C. N. Coulston. *Roman Military Equipment from the Punic Wars to the Fall of Rome*. Oxford: Oxbow Books, 2006.

Chase, Kenneth. *Firearms: A Global History to 1700*. Cambridge: Cambridge University Press, 2003.

Cipolla, Carlo. *Guns, Sails and Empire: Technological Innovation and the Early Phases of European Expansion, 1400–1700*. New York: Funk & Wagnalls, 1965.

Crosby, Alfred W. *Throwing Fire: A History of Projectile Technology*. New York: Cambridge University Press, 2002.

Ellis, John. *A Social History of the Machine Gun*. Baltimore: Johns Hopkins University Press, 1986.

Hall, Bert S. *Weapons and Warfare in Renaissance Europe: Gunpowder, Technology, and Tactics*. Baltimore: Johns Hopkins University Press, 1997.

McDougall, Walter A. *. . . The Heavens and the Earth: A Political History of the Space Age*. Baltimore: Johns Hopkins University Press, 1997.

McNeill, William H. *The Age of Gunpowder Empires, 1450–1800*. Washington. D.C.: American Historical Association, 1989.

McNeill, William H. *The Pursuit of Power: Technology, Armed Force, and Society since A.D. 1000*. Chicago: University of Chicago Press, 1982.

Rhodes, Richard. *The Making of the Atomic Bomb*. New York: Simon & Schuster, 1986.

Van Creveld, Martin. *Technology and War: From 2000 B.C. to the Present*. New York: Free Press, 1989.

Web Sites

The Cave of Lascaux
www.culture.gouv.fr/culture/arcnat/lascaux/en/
Run by the French government's Ministry of Culture, this site offers an account of the prehistoric art at Lascaux, including its development, discovery, excavation, eventual closing, a virtual tour of the cave network, and links to similar archaeological sites.

Computer History Museum
www.computerhistory.org/exhibits/
This Silicon Valley–based museum's site offers a timeline of computer history and online exhibits on the Babbage engine, computer chess, microprocessors, and a history of the Internet.

History of Agriculture
www.adbio.com/science/agri-history.htm
Narrative of the development of global agriculture from prehistoric times to the present. Contains a section specifically on agriculture and government price controls in the United States.

Museum of the History of Science, University of Oxford
www.mhs.ox.ac.uk/exhibits
Online exhibits from Oxford University's collection include "Science in Islam," a history of the drug trade, the world's largest collection of astrolabes, and "Wireless World: Marconi and the Making of Radio."

Pyramids: The Inside Story
www.pbs.org/wgbh/nova/pyramid/
Part of the *Nova* television series' *Online Adventure* sites, this page offers a history of ancient Egypt, interviews with archaeologists, and 360-degree views of the pyramids and their surroundings.

Railroad Historical
www.rrhistorical.com
A collection of links to railroad historical and technical societies, museums, histories, and models.

The Science Museum, London
www.sciencemuseum.org.uk/onlinestuff/subjects/engineering.aspx
Maintained by Britain's major science museum, this site offers illustrated slide shows about historical topics such as the rise of the factory system, the construction of the British railway network, and the use of science in warfare.

Smithsonian National Air and Space Museum
www.nasm.si.edu/research/
From the premier American museum on the subject, this site contains a database of artifacts in the Smithsonian collection, a collection of short biographies of women in aviation and space history, and an online image collection of African American pioneer aviators.

Water-Raising Machines
www.ummah.net/history/scholars/water
In the educational section of a news and religion site for the Muslim community, this page details the history of Arabic irrigation devices.

Index

Page numbers in **bold** indicate illustrations.

Dorians, 39
Douglas DC-3, 122
dragon-backbone machine, 52
drills, 100
droughts, 18, 21
dugout canoe, 47
DuPont, 116, 118
Dutch, 81, 89, 126
dynamite, 116

Earhart, Amelia, 121
Early Bird, 140
early humans, 6, 8
East Africa, 72–73. *See also* Africa
Easter Island, 15, 47
Eastman Kodak, 116, 118
Eckert, J. Presper, 132, 141
Ecuador, 68
Edison, Thomas, **114**, 115–117, 131
Edison Electric, 119
Edo, 88
Egypt
    agriculture, 11
    Assyrians, 39
    and Britain, 107
    calendars, 20
    clothing, 28
    copper, 30
    cotton, 29, 107
    dhows, 73
    elites, 107
    gnomon, 63
    Hyksos invasion, 38–39
    industrialization, 107
    iron smelting, 37
    jewelry, 28–29
    navigation, 45
    paper manufacturing, 107
    public works projects, 41
    pyramids, **25**, 26
    Sea People, 39
    shipbuilding, 107
    sugar cane, 82
    water control systems, 19–20, 54
    water house, 20
    weaving, 14, 28
    wheeled vehicles, 32
    writing, 32–33
Einstein, Albert, 131
El Paraíso, 27
electricity, 113, **114**, 115–116, 118, 123, 127,
    132, 135
electronics, 138
elites
    African kingdoms, 38
    China, 70, 92
    and civilization, 18, 34
    clocks, 64
    Egypt, 107

industrialization, 109–110
and innovation, 35
Japan, 79, 126
Latin America, 107
medieval culture, 56, 60
metallurgy, 27, 30
rice cultivation, 53
running water, 42–43
Russia, 106
Shang dynasty, 21
silk clothing, 29
textiles, 28
water control systems, 24
weapons, 31
wheeled vehicles, 27
women and empire, 48–49
writing, 34
emigration, 112
empires, 35–36, 39, 41, 43–45, 48, 50, 54, 123
energy, 111, 148
England. *See also* Britain; London
    arms race, 81
    aviation, 120
    cannons, 89, 96
    cast iron, 81, 89, 96–97
    code-breaking, 132
    conquest of, 60
    DNA (deoxyribonucleic acid), 144
    Domesday Book, 61
    jet engine, 131
    magnetism, 114
    market fairs, 67
    passenger planes, 131
    steamboats, 101
    telephone cables, 140
    windmills, 62
English Channel, 60
ENIAC (Electronic Numerical Integrator and
    Calculator), **132**
Enola Gay, 130
entrepreneurship, 92, 107
environmental damage, 105, 124, 146–148
Ephesus, 43–44
Essentials of Agriculture and Sericulture, 85
Ethiopia, 1–2, 37
Eurasia, 32, 35, 39, 41, 50, 54, 61. *See also*
    Asia
Europe. *See also* individual countries
    Age of Exploration, 75
    agriculture, 11, 56–59
    air pollution controls, 147
    broadcasting, 119
    bronze, 31
    cannons, 79–80
    chariot warfare, 38–39
    compass, 75
    computers, 142
    Dark Ages, 55–56
    early humans, 6

Langley, Samuel, 120
language, 6, 37–38
lapis lazuli, 28
laptop computers, 142
lasers, 140
lateen sail, 73, 75
lathes, 100
Latin America, 104, 107, 112
Leakey, Mary, 1
leather, 27
Lenin, Vladimir, 125
Lenoir, Etienne, 119
lentils, 13, 19
Levant, 3, 11
Li Kang, 64–65
Libya, 124
light bulb, 114–115, 117, 123
lightning rod, 113
Lilienthal, Otto, 120
limestone, 25, 37
Lincoln, Abraham, 104
Lindbergh, Charles, 121
linen, 28
Lisbon, 75
literacy, 55–56, 84
literature, 17, 49
living standards, 70, 87–88, 126–128
llamas, 12, 24, 28, 69
llanos, 82
locks, 66, 69, 73
London, 55, 89, 97, 105, 133, 147. *See also*
    England
longitude, 76
looms, 14, 95. *See also* weaving
lost-wax method, 31
Lunar Society, 91–92, 98

machine guns, 124
Machine Tractor Stations, 126
machines, 91
Machu Picchu, 69
Madagascar, 48
magazines, 122
Magellan, 75
Magic Canal, 43
magnets, 72
Mahometta, 78
mainframes, 141
maize, 12, 21, 24, 82–83
Malacca, 80
malachite, 30
malaria, 124
Malay Peninsula, 47
Malindi, 71, 73
Malta, 15
mammoths, 3, 7
Manchester, 105
Manchester-Liverpool Railway, 102
mandarins, 88

Manhattan, 140
Manhattan Project, 131
manioc, 82–83
Manuel, 80
mapmaking, 137
Marconi, Guglielmo, 118–119
Mare Nostrum, 45
Martel, Charles, 59
Maskelyne, Nevil, 76
mass production, 40, 49, 91, 120, 122–124
Massachusetts Institute of Technology, 117, 131
master farmers, 53
matchlocks, 79
mathematics, 17, 55, 69
Mauchly, John, 132, 141
Maxwell, James Clerk, 118
Mayans, 27
meat-packing, 120
Mecca, 72
mechanical engineering, 36, 48, 61, 91
medicine, 55, 56, 69
medieval culture, 58–61, 64
Mediterranean, 10, 45, 74–75, 104, 111
megaliths, 15–16
Meidum, 26
Meiji oligarchs, 126–128
Melanesia, 47
melons, 54, 82
Menes, 20
Menkaure, 26
Menlo Park, 117
merchant ships, 80
Meroë, 37
Mesa Verde, 27
Mesoamerica, 29, 33
Mesopotamia
    Assyrians, 39, 40
    civilization, rise of, 18
    copper, 30
    cotton agriculture, 29
    iron smelting, 36–37
    navigation, 45
    potter's wheel, 32
    pottery, 30
    rainfall, 21
    trade networks, 14
    warfare, 25
    water control systems, 54–55
    weaving, 14, 28
    writing, 32–33
Messerschmidt, 131
metal working, 14–15
metallurgy, 27, 30–31, 87
meteorology, 137
Mexico, 12, 20–22, 23, 27, 31, 82–83, 145, 147
microliths, 4
Microsoft, 142
microwaves, 140
Middle Ages, 65, 70

rolling mills, 97, 102
rolling stock, 112
Rome
  barbarian invasion, 54–55
  bronze, 40
  catapults, 49, 60
  fall of, 59
  naval power, 47
  navigation, 45
  nomadic warriors, 35, 54
  pilgrimages, 67
  public works projects, 41
  road building, 44–45, 66
  sundials, 63
  water wheels, 50, 61
  wheeled vehicles, 67
Roosevelt, Franklin D., 133
routers, 143
Royal Road, 43–44
Russia. *See also* Soviet Union
  atomic bombs, 134
  chariot warfare, 38–39
  early humans, 5–6
  elites, 106
  empires, 123
  gunpowder empires, 78
  horse domestication, 38–39
  hydroelectric power plants, 126
  industrialization, 106–107, 125–126
  Ivan the Terrible, 78
  and Japan, 123, 127
  living standards, 126
  middle class, 106–107, 125
  Nicholas I, 106
  Parthians, 40
  railroads, 106, 125
  Soviet Union, 125–126, 133–134, **136**,
    137–138, 141
  technologies, 125
  telegraphy, 104
  Trans-Siberian railroad, 125
rye, 11, 57, 81

saddle, 66–67
Safavid dynasty, 78
SAGE network, 141
Sahara Desert, 37–38, 67
Sahul, 7
sail, lateen, 73
sailing ships, 102
Saint Petersburg, 106
samurai, 79
San Francisco, 142
sanitation, 88
Santiago de Compostela, 67
saqiya, 50, 54
satellites, 137, 140
sawmills, 61
saws, 4, 36

scalpels, **56**
Scandinavia, 14, 58
Schall, Adam, 80–81
sciences, 54–55, 69, 111, 113, 133
scrapers, 3, 5
sculpting, 6, 26–27
sea barbarians, 109
Sea People, 39
seashells, 13
seed drill, 19
Segovia, 42
self-strengthening movement, 128
semaphores, 102–103
Senegambia, 12
servers, 143
Severn river, 96–97
sewage systems, 124
sewing machines, 100
shaduf, 19–20
Shang dynasty, 21, 29, 31, **32**, 39. *See also* China
sheep, 10–11, 19–20, 81, 144
sheepskins, 84
shelter, 5, 8, 13, 19, 88
Shi Huangdi, 39, 43–44, 66
shipbuilding
  container ships, 146
  deforestation, 96
  Dutch, 89
  Egypt, 107
  and empire, 36
  Europe, 74–75, 89
  India, 107–108
  Japan, 127, 138
  Mediterranean, 74
  Portugal, 74–75, 89
  Song dynasty, 71–72
  Spain, 74–75, 89
  steam engines, 112
  steamboats, 100–103
  steel manufacturing, 113
  Turkey, 87
  warfare, 45, **46**, 47
Shippingport, 135
Shockley, William, 137
shoguns, 88
shortwave radio, 119
Siberia, 6, 8, 58, 78
Sichuan, 43, 53
Sicily, 45, 82
sickles, 8, 14, 65
side-wheeler, **101**
siege machines, 40, 59, 77–78
Siemens, 116, 127
Siemens, Werner von, 114
Siemens-Martin, 112, **113**
signals, 112
silk, 21, 28–29, 49, 51, 53, 63
Silk Road, 44, 67
silver, 15, 28, 31, 107

and population, 92
research laboratory, 117–118
Russia, 125
and science, 113
steam engines, 99
Stone Age, 7
and tariffs, 146
time measurement, 63–64
and trade, 146
twentieth century, 125
United States, 125, 138
warfare, 40, 50, 130
Western world, 138
World War I, 130
World War II, 147
Teflon, 134
Teheran, 147
Telefunken, 118
telegraphy, 103–104, 114
telephones, **115**, 118, 139–140, 143
television, 137–138, 147
Telstar, 137
Temple of the Moon, 27
Temple of the Sun, 27
temples, 27, 68–69
tenant farmers, 53, 60, 92
teosinte, 12
Teotihuacán, 22, **23**, 27
terraces, 52
Tesla, Nicola, 116
test ban treaty, 134
test-tube babies, 144
Texas Instruments, 138
textiles
 Africa, 38
 agriculture, 27–28
 Britain, 92
 calico, 94
 chemical pollution, 105
 child labor, 95–96
 China, 69, 92
 Greece, 28
 industrialization, 106, 108
 manufacturing of, 113
 Mediterranean cultures, 28
 Neolithic Revolution, 14
 Peru, 28
 Russian manufacturing, 106
 silk, 51
 spinning wheels, 62–63
 Teotihuacán, 61
 waterwheels, 61
 women's work, 14
Three Mile Island, 135
three-field rotation, 56–58
threshers, 117, 126
Thucydides, 46
Tibet, 81
Tigris-Euphrates Valley, 10, 18

Tikal, 27
timber, 20
tin, 31, 36, 98, 113
Titanic, 112
Tiwanaku, 24
TNT, 134
tobacco, 82
toilet paper, 84
Tokugawa Ieyasu, 79
Tokyo Bay, 79, 126
tomatoes, 12, 82
tongs, 36
torsion catapults, 61
Toshiba, 127
Totsuko (Tokyo Telecommunications
 Engineering Company), 139
trade networks
 and agriculture, 58, 70
 Arabs, 54–55
 Black Sea, 45
 Britain, 14, 92
 China, 73
 copper, 30
 Europe, 67, 74
 Indian Ocean, 80
 Indus River Valley, 20
 and industrialization, 128
 Latin America, 107
 Mediterranean, 45
 Mesopotamia, 14
 Mexico, 21
 Middle East, 66
 Muslims, 38
 and navigation, 74
 Neolithic Revolution, 14
 obsidian, 13
 Portugal, 80–81
 pottery, 30
 road building, 43–44
 South America, 24
 Spain, 14
 spices, 74
 steamships, 112
 and technology, 146
 tin ore, 31
 trade winds, 75
 and transportation, 65, 67
trade winds, 75
transformers, 116
transistors, 137–138, 140–141, 147
transportation. *See also* railroads
 automobiles, 124, 147
 aviation, 120–122
 China, 66
 container ships, 146
 electricity, 116
 emigration, 112
 and empire, 36
 marine, 45–48

The New Oxford World History

Forthcoming Titles

CPSIA information can be obtained at www.ICGtesting.com
Printed in the USA
BVOW012033181212

308508BV00003B/9/P